Upgrading Physics Education to Meet the Needs of Society

Maurício Pietrocola
Editor

Upgrading Physics Education to Meet the Needs of Society

Springer

Editor
Maurício Pietrocola
Faculty of Education
Sao Paulo, Brazil

ISBN 978-3-319-96162-0 ISBN 978-3-319-96163-7 (eBook)
https://doi.org/10.1007/978-3-319-96163-7

Library of Congress Control Number: 2019930721

This Springer imprint is published by the registered company Springer Nature Switzerland AG
The registered company address is: Gewerbestrasse 11, 6330 Cham, Switzerland

Foreword

This book is an initiative, recently promoted by GIREP by means of an agreement with Springer to contribute in physics education research and praxis with a series of publications. It contains selected contributions on the topic "Contemporary Science Education and Challenges in the Present Society: Perspectives in Physics Teaching and Learning." By a peer review process involving 854 reviews carried out by 149 referees, we selected the papers published in this volume among 307 contributions we had already selected for the presentations in the Second World Conference on Physics Education (2nd WCPE). We have to thank Maurício Pietrocola for editing the book and leading the peer review process after the organization of the 2nd WCPE and Claudia Haagen Schuetzenhoefer for managing the editorial process for GIREP.

GIREP is an international membership organization founded in 1966, open to academics, teachers, curriculum developers, and all other stakeholders with the concern to improve physics teaching and learning by means of physics education research (PER), innovative experimentation in physics teaching/learning, innovative materials and methods, suggestions for stakeholders, international cooperation in conferences, seminars, and selected paper books. In the past 50 years, 1400 physics education researchers, teachers, and other specialists of 72 countries shared their common problems, studies, results, and experience in GIREP. Out of researchers and teachers scattered in faraway schools and colleges and universities, GIREP created a community. From 1967 to 2016, GIREP organized 32 international conferences in cooperation with EPS, ICPE, MPTL, and UNESCO.

GIREP's main focus is to connect research and teaching from primary to university level. This is motivated by the following reflections. The quality of teaching is determined by cross-fertilization of research and praxis. Teacher education and teacher professional development need a connection with physics education research, and we experiment that this is not extensively done in the different countries. Continuity in vertical and in transversal perspective is very important. A large society engagement is needed to promote scientific learning, policies for teacher education, and quality in teaching and learning. We mainly focus on content research, because learning is content specific; subject matter and pedagogy are not enough for teacher professional development: there is a wide need on didactics on subject matter. Our goals are to improve practice by means of research (PER), to promote content structure research, and to explore teaching/learning processes for new topics conceptual learning and lab work, methods, strategies, and tools' role in learning. Active research lines in GIREP represented in this volume are different: content-oriented theory, students' conceptions/reasoning, students' learning pathway and processes, developing content-specific tests, role of approaches, concepts, contexts, motivation for learning on specific topics, individuate conceptual profiles and parallel conceptions, variation theory, design-based research, curricular research, conceptual profile, learning progression, and teacher education.

In the last 5 years, GIREP formalized the cooperation with the following international bodies, signing agreements: American Association for Physics Teaching (AAPT), American Physical Society (APS), Education Division of European Physical Society (EPS-ED), European Science Education Research Association (ESERA), Inter-American Conference on Physics Education (IACPE), International Commission on Physics Education (ICPE), interAsian Scientific Education Research (iSER), Latin American Physics Education Network (LAPEN), Multimedia Physics Teaching Learning (MPTL), International Association of Physics Students (IAPS), and Japanese Physics Education Society (PESJ). In this framework, the exchange of research results and experience gain an open high scientific value for each GIREP conference and find a common working area initiative in the WCPE 4-year conferences.

The world of physics education research does not have a wide range of journals in which to publish and compare the results of their work. Making available in a book the best research contributions in physics education and significant teaching interventions presented at the 4-year World Conference on Physics Education, it seemed to us the best way to serve the research community and the stakeholder in the field of physics education. We hope this contribution will be useful to this community and to teachers in general. Comments and suggestions will always be welcome in order to better realize the mentioned GIREP objectives.

GIREP Committee, Physics and Math
Section of DCFA University of Udine
Udine, Italy

Marisa Michelini

Contents

Chapter 1
How Should We Teach Physics Today?

Maurício Pietrocola

Abstract This question is being asked by physicists around the world in the face of the global societal changes that have taken place over the last 50 years. Those same five decades have seen the publication of the original version of the current best-selling book in the world for teaching physics at the university level, as well as many updated editions. First released in 1960, Halliday and Resnick's *Physics for Students of Science and Engineering*—now in its 10th edition and known as *Fundamentals of Physics*—has shaped the teaching of basic physics at the university level and strongly influenced physics in secondary education. The authors' initial intentions clearly involved creating a community of science and engineering researchers capable of developing and incorporating basic and technological knowledge for the benefit of postwar industrial society.

This question is being asked by physicists around the world in the face of the global societal changes that have taken place over the last 50 years. Those same five decades have seen the publication of the original version of the current best-selling book in the world for teaching physics at the university level, as well as many updated editions. First released in 1960, Halliday and Resnick's *Physics for Students of Science and Engineering*—now in its 10th edition and known as *Fundamentals of Physics*—has shaped the teaching of basic physics at the university level and strongly influenced physics in secondary education. The authors' initial intentions clearly involved creating a community of science and engineering researchers capable of developing and incorporating basic and technological knowledge for the benefit of postwar industrial society.

Although the Halliday and Resnick collection continues to be the main reference for physics teachers around the world today, the early twenty-first century is very different from the late 1950s. Today's society presents individuals with difficult dilemmas. As science and technology progress, we find ourselves on the frontiers of knowledge and capabilities so complex and extraordinary that absolutely no one is

M. Pietrocola (✉)
Faculty of Education, University of São Paulo, São Paulo, Brazil
e-mail: mpietro@usp.br

M. Pietrocola (ed.), *Upgrading Physics Education to Meet the Needs of Society*,
https://doi.org/10.1007/978-3-319-96163-7_1

able to fully understand their contours. At the same time, this unprecedented context generates a diversity of possible futures about which we need to be informed in order to develop opinions and make crucial decisions.

In general, two fundamental transformations are affecting the lives of people today. Both are connected with the growing influence of science and technology, though not completely determined thereby: the end of nature and the end of tradition.

The world of nature that humankind has learned to deal with and to know since the very beginning of our existence no longer exists. Over the past 75 years, humanity began to be less concerned with what nature could do to us and to worry more about what we had done to nature. The awareness that the environment in which we live had begun to degrade is something that dates back at least to the 1980s. It has become increasingly clear that human settlements are not surrounded by nature; on the contrary, nature has been surrounded by humanity. This has occurred to such an extent that our wild resources—nature preserves, forests, wetlands, rivers, springs, oceans, and so on—have become objects of action for groups defending nature. This transition is fundamental to our entry into what Beck (1988) defined as a risk society. We are now a post-nature society, reflecting and experiencing how science and technology have transformed nature into techno-nature.

In this modified society, the earlier societal mode of danger-security pair, in which tradition took center stage, has been replaced by risk-confidence. This new status may seem to imply a zone of semantic overlap between danger and risk, but these notions belong to very different worldviews. The idea of risk is connected with aspirations toward control and in particular with the idea of controlling the future—a relatively recent concern in human history which probably stems from the end of the Modern Age. Social risk emerges as a factor in societies when they start to worry about the future and to seek to guarantee its success and safety. European explorers in the fifteenth century, who sought to minimize risk on overseas trips by means of techniques and knowledge, were some of the first to implement this style of risk management. From this novel perspective, risk entails both negative and positive connotations. On the one hand, it is associated with the idea of avoiding an unwanted result. But its positive side resides in the ability to take initiative in the face of a problematic future. Thus, risk-confidence societies must develop their citizens' abilities vis-à-vis managing risk.

And what are the risks today? Paradoxically, the current need for the continual production of knowledge and technology brings risks of various kinds: environmental degradation, increasing poverty, widening inequalities, exclusion of minorities, etc.

It is important to highlight certain important developments in the science-society context as constitutively related to these risks:

- Over the last two centuries, science and technology developed in such a way that science became the cornerstone of the Western tradition.

- Scientific knowledge was once seen by most (and is still seen by some) as capable of overcoming previous traditions but has become a given, an authority certain, and secure in its own right.
- Because as science and technology became more complex, they became more and more distant from people's lives; people began to respect them as a means of generating security.
- In the absence of access to learning and information that would allow them to form their own opinions—and to calculate and manage their own risks—laypeople sought the opinions of the experts (scientists and engineers).
- We thus now need ways to inculcate a much more dialogic and engaged relationship with science and technology in students—and in citizens in general.

Within this context of questioning which types of scientific knowledge should be taught, the contents that emerge as key are those aimed at analyzing the role and importance of such knowledge to the basic formation of a social conscience in the individual. The field of physics—influencing as it does students' perceptions of the natural world—has much to offer in this process

The report to the European Commission of the expert group on science education (2015) affirms:

> We need science to inform policy, objectively. We need science to inform citizens and politicians in a trustworthy and accessible way. We need to make decisions together—rather than from polarised positions—and to take responsibility for those decisions, based on sound scientific evidence. (p. 5)

These kinds of needs cannot be met if we think of how physics has been taught and learned. In this sense:

> Science education research, innovation and practices must become more responsive to the needs and ambitions of society and reflect its values. (p. 6)

For this we must turn to three basic questions: Why do we teach? What do we teach? How do we teach?

The present book is part of the movement to respond to these questions from the physics teaching research point of view.

The chapters of part 1 of the work take the influence of science and technology in the means of teaching physics. Computers, technology for teaching, and contemporary content worked in school are the focus of the chapters.

Ian Lawrence states that computers can be usefully thought of as representation tools. Many demonstrated difficulties in learning physics depend on re-representing the world to yourself: imagining it as other than it appears and then reasoning with that new representation, to develop new expectations about the lived-in world. To keep physics live in classrooms during this process requires the most responsive and adaptable tool we can lay our hands on, to encourage teachers to do physics with children. Rather careful thinking about matching the desirable affordances and resistances present in the practices enabled by any tool to the existing physics curriculum suggests casting the net more widely than numerical integration of differential equations. This paper draws on a number of years of working with computational modelling tools with teachers and with pupils (8–18), as well as significant work in constructing teaching sequences and supporting representations in the Supporting Physics Teaching initiative and Advancing Physics (both supported by the Institute of Physics). The foci are on exploiting flexible diagrammatic representations without being able to draw and on evolving responsive representations to support developing ideas during teaching sequences while seeking also to exploit the new enthusiasm for coding in a culturally valuable way and on keeping the implementation straightforward enough that teachers might be persuaded to use it.

Tom Ellermeijer e Trinh-Ba Tran willing to answer the questions "How to make physics education more challenging, relevant, and attractive for our high school students? How to stimulate the development of creative thinking, problem-solving, and other higher cognitive skills?" In many countries governments like to stimulate science and technology in schools but in this direction, STEM (or STEAM), IBSE, and MINT (Germany) are the more recent alphabet soup acronyms. Can technology applied in physics education bring us closer to the desired goals? Clearly it has been demonstrated that technology can help make physics education more relevant, more linked to real life, and more authentic and can increase the opportunities for own investigations by the students. So it really has an added value and not just provides another way of teaching the same. This is known for decades but still applied at a relatively small scale. They will present several examples of the use of measurements with sensors, video measurements, and modelling demonstrating these benefits.

Marcia Begalli and Uta Bilow present issues from activities where research institutes and universities around the world invite students and their teachers for a daylong program to experience life at the forefront of basic research. These International Masterclasses (www.physicsmasterclasses.org) give students the opportunity to be particle physicists for a day by analyzing real data from CERN's Large Hadron Collider (LHC). The project attracts each year more than 13,000 high school students from 46 countries. In the International Masterclasses, high school students work with real data collected by the experiments at the LHC. The program bridges the gap between science education at school and modern scientific research. Participants can explore the fundamental forces and building blocks of nature and are informed about the new age of exciting discoveries in particle physics, e.g., the *Discovery of the Higgs Boson*. Moreover, they can actively take part in cutting-edge research and improve their understanding in science and the scientific research process. The program offers authentic experience and adds valuable experiences to physics education at school, thus stimulating the students' interest in science.

In part 2, the privileged topics are evaluations about established methods of teaching physics, which are the peer instruction method and the simple experiments.

Hideo Nitta develops a few mathematical models of learning that were developed previously. Then he presents his mathematical model that describes dynamics of the response of students in peer instruction (PI). In this model, for evaluating the effectiveness of each question for PI, the "peer instruction efficiency (PIE)" is introduced in analogy with the Hake gain. It is shown that, in the simplest approximation, PIE becomes proportional to the relative number of students answering correctly before discussion. The mathematical model is applied to introductory physics courses at a university and physics classes at a high school. It is found that overall practical data of PIE moderately agrees with theory. Application of PIE to data analysis is also discussed.

Leos Dvorak develops considerations about simple experiments, being cheap, requiring just a short time, and often providing interesting results; simple experiments have been used in physics classes for a long time. However, how is it nowadays? Can they compete with what ICT, modern technologies, and ubiquitous sophisticated gadgets provide for us and our students in more and more attractive forms and formerly unimaginable range and quality? Are not simple experiments with a few straws, skewers, and threads obsolete and even ridiculous in the age of the Internet offering tons of applets, virtual labs, and latest achievements of physics at multimillion-dollar international facilities with one click? The purpose of the talk is to show that simple experiments "are not dead" and can be much more that qualitative toy experiments and their potential are greater than just to generate "small wow" reaction in students. Of course, this statement is something most physics teachers and educators would probably agree with; we like such experiments and the fact that they are useful belongs to internal beliefs of most of us. However, it should not be just a blind belief. Therefore, the aim of the lecture will be to support this claim by concrete examples (some of them hopefully at least partially new for the audience). So, we will try to "add ammunition" to our conviction that simple experiments can really have their firm place and good perspectives even in physics education in times to come.

In recent years, the subject of innovation has been a recurring theme in the international literature on science education research. Perhaps because of the large amount of update/renovation projects for school curricula in the last few years, this focus has been adopted by many researchers in the field, making it a point of study. These works are normally related to projects aimed at introducing and evaluating the impact of curricular innovation (Pinto 2002, 2005; Ogborn 2005; Mansour 2010).

In part 3 of the work, the focus becomes the results of research-based alternatives to traditional physics.

Jenaro Guisasola provides an overview of trends with regard to different methodologies of instruction and analysis of students' learning. The chapter aims to describe and discuss teaching approaches and students' achievement on specific topics of the curriculum at university level. At university level, scientific-technological education should support a diverse student population where actually using knowledge, not just memorizing it, is becoming more important. He discusses and compares teaching approach frameworks and their features across different characteristics, such as transformation of the content, explicit monitoring of students' learning, and evaluation. He draws different lines of teaching approaches.

Cristiano Mattos trabalja na perspectiva de prospectar alternativas ao ensino tradicional. Your main purpose is to localize alternative teaching proposals in the education's activity chain, since higher education activity is part of a larger educational system that connects basic school to productive working life.

In the same direction, Limiñana, R., Menargues, A., and Rosa, S. will present the development of a teaching/learning sequence based on a structure of problem-solving that generates a tentative environment, where students and teachers have to plan a possible strategy to advance/solve the problem, carry out this plan, and analyze results. He will analyze the specific topic of latitude and longitude.

Marisa Michelini and *Alberto Stefanel* present "Innovation in Physics Teaching/Learning for the formative success in introductory physics for Bio-Area degrees: the case of fluids." Research-based intervention modules are studied in the last 2 years, for degrees in the University of Udine. The main aspects to be faced are:

(A) To redesign the way in which physics is offered so that its role can be recognized in the specific subject matter characterizing the degree: turning the ways in which physics is approached, changing the role of each topical areas, and individuating specific applications of physics in the professional field of the degree
(B) To offer instruments and methods building a physics competence in different fields
(C) To individuate strategies able to produce an active role of students in learning physics and to give them the opportunity for an appropriation of the applied physics methodologies
(D) To support students learning in multitasking ways by means of ICT tools, of lab activities, and of problem-solving and step-by-step evaluation of learning outcomes

Genaro Zavala presents "the design of problems based on cognitive scaffolding to teach physics." He has been working on the design of problems based on cognitive scaffolding to teach physics. These problems are designed to be used in almost any setting since no equipment is needed. Students work in collaborative groups of three or four students each. The design consists on transforming a traditional problem to a tutorial-format problem which takes the student through scientific reasoning steps to build concepts, that is, cognitive scaffolding. In this contribution some examples will be presented, and results of reasoning of students will be analyzed.

Mila Kryjevskaia will deal with "Examining the Relationships Among Intuition, Reasoning, and Conceptual Understanding in Physics." In an ongoing project focusing on student reasoning in physics, she has been developing and applying various methodologies that allow her to disentangle reasoning, intuition, and conceptual understanding in physics. The dual process theory is used to account for the observed patterns in student responses. Data from introductory physics courses will be presented, and implications for instruction will be discussed.

We suggest that there has been progress but that more work is needed toward identifying the effectiveness of the approaches in different countries with similar contexts and curriculum. The products of the innovation must be reproducible, as is not often the case, to constitute a reasonable foundation of accepted didactical material.

In her chapter Laurence Viennot states that critical thinking is unanimously presented as of central importance to science teaching. But the present focus on competencies observed in many countries correlates to the reduced conceptual structuring of resources that are commonly used in teaching. A crucial question then arises: is it fruitful to envision conceptual and critical developments separately? In operational terms, can critical thinking be fostered in students without conceptual structuring? Based on an epistemological framework stressing the pivotal role of a search for coherence in science, the content to be taught will be referred to a dialogue between a thorough content analysis and what we know of students' prescientific ideas, also keeping in mind the striking persistence of teaching rituals. The type of critical attitude considered here consists in questioning explanations that would be inconsistent or very incomplete, in the search for intellectual satisfaction. This chapter presents a brief synthesis of some recent investigations bearing on the co-development of conceptual understanding and critical attitude in university students. In characterizing students' responses when confronted to various explanations of a physical phenomenon, these studies bring to bear conceptual markers as well as meta-cognitive, affective, and critical indicators. Some profiles of co-development will be characterized, including "delayed critique" and "expert anesthesia of judgment." The results strongly suggest that to disregard the objective of conceptual structuring is counterproductive for the development of students' critical attitude. Through these exploratory studies, it appears that the conditions in which students can begin to search for coherence—whether in pursuit of conceptual understanding or to activate their critical potential—constitute a crucial objective for further research.

In part 4, the book turns to an emergent subject in all educational contexts, i.e., the necessity to encompass the teaching for and from diversity and difference among persons, nations, and knowledges.

Antonia Candela and Johanna Rey provide ethnographic descriptions and analyses of interviews with indigenous and Afro-Colombian teachers and of some discursive interactions with their students in primary-school classrooms in underserved communities. At those contexts they mobilize their local community knowledge for science lessons. We analyzed the teachers' purpose in incorporating indigenous and Afro knowledge in teaching science and how these different knowledge systems work in the interaction. These teachers' and students' co-constructions modify and enhance the official science curriculum with forms of resistance to the scientific myth of only one universal truth about physical phenomena. This resistance is based on the strength of their collective identity constructs as well as their connection with and respect toward nature. These kinds of studies are relevant references for a culturally sensitive science curriculum development.

The goal of *Katemari Rosa* chapter is to discuss academic climate for underrepresented groups in Brazilian physics departments. The conversation stems from looking at hate crimes happening worldwide and asking whether this hateful environment of society at large affects academic institutions. Would sexism, LGBTphobia, and racism be present in physics classrooms? Could hate speech or behavior, somehow, affect physics teaching and learning? Grounded on feminist perspectives, theories of identity, and critical race theory, the paper looks into diversity and physics education by examining the situation of race, gender, and sexual minorities in physics. Specifically, it takes on hashtag activism to analyze the experiences of students from underrepresented groups in science. The site of research is social media and the narratives produced by #MyTeacherSaid in Brazil, which was a hashtag used to reveal aggressive comments professors make to students. Results show that analyzing activism through social media can be helpful for unveiling oppressive environments in academia. Specifically, this study shows there is an oppressive climate for gender, racial, and sexual minorities of Brazilian students in STEM. The comments range from subtle but harmful comments loaded with gender and race stereotypes to open threats to students. Finally, the paper urges for a change within physics education research community to include intersectional approaches that take into account race, gender, and sexuality so that we can better understand the teaching and learning of physics, in addition to providing resources to help making more inclusive STEM environments.

Tanja Tajmel's chapter deals with "diversity in physics education" from a theoretical and discourse-analytical point of view. The discourse on "diversity" is being examined from different perspectives. Due to different motives in promoting diversity—utilitarian as well as emancipatory ones—a critical awareness of physics teachers and education researchers regarding diversity becomes increasingly relevant. The common conceptual delineation of diversity, especially through categorizing individuals by certain characteristics, bears the risk of "othering" and discrimination. In this contribution, the human rights perspective is highlighted as an approach toward a critical understanding of diversity in physics education.

References

Mansour, N. (2010). Impact of the Knowledge and Beliefs of Egyptian Science Teachers in Integrating a STS Based Curriculum: A Sociocultural Perspective. Journal of Science Teacher Education, 21(5), 513–534.

Ogborn, J. (2005). Introducing relativity: less may be more. Physics Education, 40(3), 213–222. https://doi.org/10.1088/0031-9120/40/3/001

Pinto, R. (2002). Introduction to the Science Teacher Training in an Information Society (STTIS) project. Int. J. Sci. Educ, 24(3), 227–234.

Pinto, R. (2005). Introducing curriculum innovations in science: Identifying teachers' transformations and the design of related teacher education. Science Education, 89(1), 1–12. https://doi.org/10.1002/(ISSN)1098-237X

Part I
Science and Technology in the Means of Teaching Physics

Chapter 2
Dialogic Development of Children's Ideas Using Computation in the Classroom: Keeping it Simple

Ian Lawrence

Abstract Computers can be usefully thought of as representation tools. Many demonstrated difficulties in learning physics depend on re-representing the world to yourself: imagining it as other than it appears and then reasoning with that new representation, to develop new expectations about the lived-in world. To keep physics live in classrooms during this process requires the most responsive and adaptable tool we can lay our hands on, to encourage teachers to do physics with children. Rather careful thinking about matching the desirable affordances and resistances present in the practices enabled by any tool to the existing physics curriculum suggests casting the net more widely than numerical integration of differential equations. This paper draws on a number of years of working with computational modelling tools with teachers and with pupils (8–18), as well as significant work in constructing teaching sequences and supporting representations in the Supporting Physics Teaching initiative and Advancing Physics (both supported by the Institute of Physics). The foci are on exploiting flexible diagrammatic representations without being able to draw and on evolving responsive representations to support developing ideas during teaching sequences whilst seeking also to exploit the new enthusiasm for coding in a culturally valuable way and on keeping the implementation straightforward enough that teachers might be persuaded to use it.

2.1 A Place for Diagrams in Learning Physics

Discussions between teachers and students have been studied and extensively theorised over the past few decades in ways that have had impact on practice and on reflections about practice. Some work has also been done on developing drawn representations and some work on what it would take to make computational modelling possible with younger children. I think it would be fair to say that these

I. Lawrence (✉)
Institute of Physics, London, UK
e-mail: Ian.Lawrence@physics.org

M. Pietrocola (ed.), *Upgrading Physics Education to Meet the Needs of Society*,
https://doi.org/10.1007/978-3-319-96163-7_2

have not challenged the hegemony of speech, either in research output or as what is seem on classroom walls, at least in the UK, as ‘topic words’. In spite of those who have theorised about communication in science classrooms, writing about the nature of explanation in the classroom, research, examining and classroom practice has tended to emphasise words as the primary medium for expressing ideas in physics.

The use of all three media—words, diagrams and computational models—in teaching and learning physics is concerned with developing and using representations to reason about situations or processes, whether exploring existing representations or shaping your own. Reasoning is possible with either exploratory or expressive use of representations.

Delineating this division between exploratory and expressive use of a (modelling) medium was a significant outcome of the London Mental Models group that did so much to establish the possibilities of computational modelling with younger children.

In classrooms, the different communicative modes described in studies of dialogic conversations point to a similar difference between exchanges used to elucidate expressions from students and other exchanges used to explore the ideas of teachers.

Diagrams are a third kind of medium for reasoning but appear as less plastic than words, being more difficult to construct and more difficult to adapt. The skill of a hand-constructed diagram is not something that seems to be as well practised as the skill of a well-constructed sentence. And many diagrams in physics are highly compressed representations, adopting any inventions that encapsulate many kinds of knowledge, both tacit and declarative.

Perhaps because we expect language to be more flexibly interpreted, we are better at filling in the gaps with language than with diagrams, especially highly encoded or compressive diagrams, that use many conventions. The lack of plasticity affects the interpretation, as the communicative act is more constrained than in the case of language: usually in the direction of requiring more commitment on the part of the person creating the attempted communicative act. An example may help. There is more precision associated with this diagram than with the statement ‘there is a force exerted on the mass’. In part this is because of the conventions associated with the diagrams, but it is also the case that constructing the diagram requires more decisions to be made by the constructor.

But there is a pedagogic danger in the propensity and ability to interpret what is ‘missing’ in linguistic acts: the filling-in carried out by teacher and children often leaves them at cross-purposes, both ‘hearing’ what they want to. Both parties adapt the inputs to fit their own ideas of the purposes of the transaction and of the ideas communicated by the transaction. This is an amalgam of constructivism and the asymmetric relationships of classroom behaviours. The plasticity of words impedes the clear communication of the ideas in physics.

More exploratory and expressive use of diagrams in classrooms, thus, seems likely to reduce this mismatch, given the greater rigidity of the diagram as a communicative tool. There are trade-offs, implicit in the encoding and decoding of diagrams, that any pedagogy that advocates their use will have to work on, but it seems that the gains from such an approach make it worth exploring.

The use of computational models in classrooms can make the rendering of the ideas being expressed or explored even more rigid, forcing a more explicit commitment of what the author intends. However this use is currently a minority interest, in spite of decades of encouragement and exploratory studies. Again the potential for gain seems substantial, but so far access to this potential has not been widely or evenly distributed, as the practice seems to happen in only a few classrooms.

2.2 Perspectives on Research and Practice

Words: A Common Probe

Formative assessment is a part of successful classrooms, and often this relies on expository writing as a probe. Children write, and teachers try to work out from their writing what they are thinking and how that can be worked on so that they can think in about physics more helpfully. All too often this writing is to satisfy external goals: it is not about the child representing and reasoning in order to figure out what is going on. So there is often an element of trying to guess what the teacher wants in the exercise. This, together with the plasticity of interpretation and the often extensive inferences about the stability of ideas revealed by these words, suggests that at the least this kind of probing might usefully be supplemented. If the idea is put to work, then we might find out more about how robust it is and how widely applied. But still there are limits to the medium, partly driven by the plasticity of the mode of expression, and partly by the nature of physics, which has adopted the quantitative route to rigour. Words are rarely sufficient—hence the unsuitability of 'What do you mean by?' as a probe.

Ideas, not words, are the real target, and we do not really know what an idea means until I see what I can do with it: ideas do not function in splendid isolation: you need to connect them up, and in particular connect them up to the lived-in world, to see what they really mean. This is a form of triangulation, of assembling different multiple perspectives on an idea by using different illuminating probes. Many research papers focus on streams of words, and this focus on language has perhaps supported teacher dialogue which so often, perhaps particularly with the harder ideas such as 'energy', seeks to settle on the correct form of words as the arbiter for what is correct and to be effectively transmitted. Words are useful, but they are not everything. For physics, which has essential connections to the lived-in world and is therefore intrinsically multimodal, the quote misattributed to E.M. Forster ('How do I know what I think until I see what I write?') is necessary, but not sufficient.

However the most common formative assessment pattern is a series of questions and answers—usually teacher's questions and children's answers. If anything this is even more subject to flexible reinterpretation than written words. Whereas the underdetermination of meaning by syllables uttered is an advantage in supporting everyday speech, this plasticity is often unhelpful in exploring understanding and misunderstanding, because there is simply too much flexibility in interpretation available to both participants in the communicative act. This warping of meaning can still be seen in action even if there are serious attempts probing the meanings—itself a hard job in busy classrooms. And the difficulties in communication persist across both expressive transactions, teacher and children working together finding out what children think, and exploratory transactions, teacher and children working together finding out what the a canonical view is.

There is also the difficulty of knowing whether the thinking is final or provisional or more likely some superposition of the two. To engage in dialogue is partly to work out what you think. There is more than a kernel of truth in the aphorism: 'How can I tell what I think till I see what I say?'

In the light of this, I think there is a case to be made for more exploratory and expressive use of diagrams, whether dynamic or static, to increase the range and variety of evidence on which we base our understanding of the ideas the children are deploying and developing. Whereas there is some work on developing representations towards the canonical (Tytler et al. 2013), there seems to be a dearth of tools that allow children and teachers to codevelop diagrams. The tools should provide some assistance and some prosthetic building blocks so that we do not need to start from just a pencil and a blank canvas. Just as words are deployed to package up collections of conventions and understandings, however particularly, so elements of diagrams carry centuries of refinement in thinking about depicting situations of processes. Consider the simple (obvious?) act of representing the force of gravity acting on an object. A diagram is rather unambiguous, and perhaps straightforward to read, after some practice (Fig. 2.1).

However saying the precisely the same thing in words takes a lot of them and requires significant explicit commitment to precision that is implicit in the diagram, being encapsulated in elements of the diagram. This kind of assisted disambiguation has, I think, considerable potential to encourage commitment and clarity in communicative acts.

2.3 Developing Sketching and Drawing

Words are commonly used and reused as an understanding develops in classrooms, adapted in both meaning and context as they are used by teachers and children. They're relatively easy to mix and match into new sentences, in which meaning and understanding evolve, whether spoken or written. There are not only many different tools for creating and rearranging written words, from pencil, paper and eraser to the

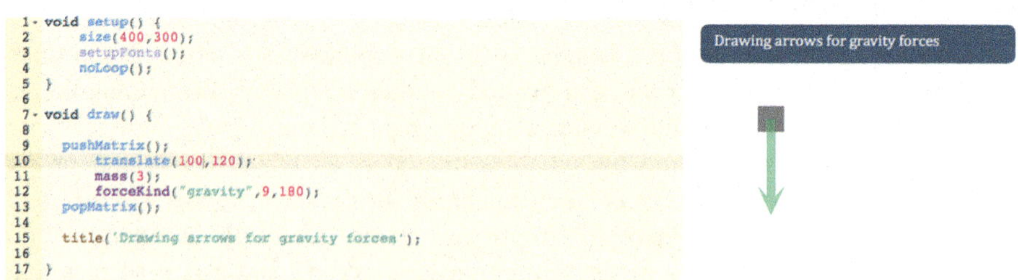

Fig. 2.1 A gravity arrow

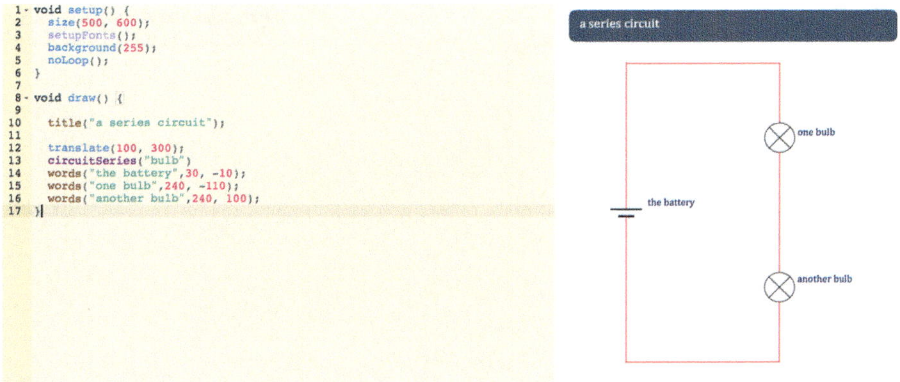

Fig. 2.2 A circuit with series connections, simply drawn

many varieties of text editor on a smartphone, but it is also the case that cultural expectations and norms firmly encourage people to acquire a competence with such tools. The inability to write grammatically is widely considered an impairment. By contrast, the inability to express yourself well using diagrams is less valued (Fig. 2.2).

Diagrammatic representations are not so often developed in classrooms: contrasting this with the affordances available for such developments in words reveals several possible reasons. Firstly elements of diagrams are less easily used and reused in such ways, when compared with the simplicity in mixing and matching words. There is, for a start, no equivalent to the enforced linear structure of the sentence (or the parallel linear interpretative framework as a result of an utterance in time, if the communication is oral). This structural freedom in diagrams adds an extra burden in both authoring and interpreting diagrams. But it may also be partly because the tools to rearrange elements of existing diagrams are somewhat more complex than text processors and partly because communicating with well-formed diagrams is less widely valued than communicating with well-formed sentences. There seems to be less of a cultural or educational imperative to develop this competence.

Yet reasoning with diagrams is a rich resource in physics: one only needs to look at the space-time diagram in relativity, the Feynman diagram in quantum mechanics and the free-body diagram in Newtonian mechanics. All three encapsulate knowledge about the domain, and manipulating the representations guides methods of reasoning about that domain.

In the educational sphere, several approaches have been made. One interesting example is to explore a set of particular geometrical interpretations in the relationships between electrical measures in resistive circuits: the AVOW diagrams. However successful, this has not been exploitable in other domains, and so there is a real question about the feedback on investment. There is a reasonable case that learning

Fig. 2.3 A simple circuit

to work with these diagrams enables children to reason successfully about a restricted class of resistive circuits. There is a question about the generalisability of the approach, as it relies on simple geometrical interpretation of relationships: it is difficult to see how this would work even in the case of the simplest non-linear relationships. This may explain why the work has not spread to other domains.

Another recent approach has explored children reworking their own representations, getting a better understanding of the value of the canonical representations. However this starts at a very low level, with the equivalent of a pencil and paper, but no words. Everything has to be built from the simplest possible operations, and there seem to be no building blocks, which to adapt and remix to construct their own diagrams.

Here I am after a meso-level, incorporating some culturally valued attributes into the elements of the diagram but allowing these to be assembled in ways that enable a degree of shaping of the communication by the teachers and children in a classroom (Figs. 2.3, 2.4, and 2.5).

Sharing an adaptable diagram is as easy as sharing the few lines of code, and changing the diagram is simple, after making the investment of time to find out how. There are inevitably questions about investment of time and trade-off between what is gained and lost: probably only pervasive use of such diagrams will tip the balance in favour of use. And diagrams are no more 'self-documenting' or self-evident than words.

On Making Adaptable Diagrams a Part of Classroom Discourse

The idea is that a kind of structured drawing can enter the classroom conversations, as a partner. Here is a connected series of diagrams, complete with the code that generates the diagram. It should not be imagined that these are all to be deployed in a single lesson, or even in adjacent lessons, but rather that they illustrate the way in

Fig. 2.4 A simple circuit, labelled

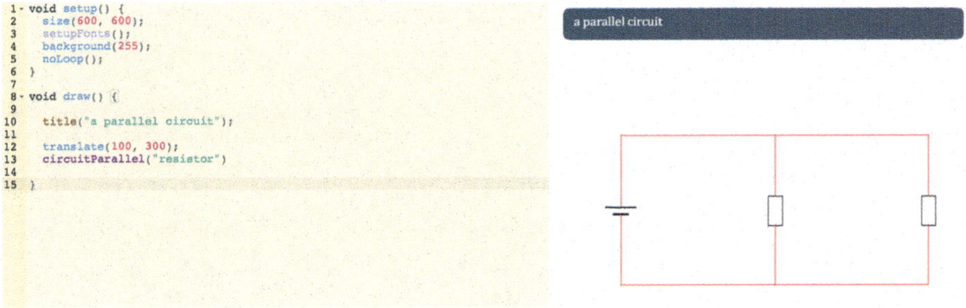

Fig. 2.5 A circuit with parallel connections, from a simple change in the code

which such a technology can encourage the expressive and exploratory use of diagrams (Figs. 2.6, 2.7, and 2.8).

A particular concern is to exploit the idea in the processing language of a 'sketch': the code remains adaptable, and one should expect to iterate the diagram, exploring your expressions and using both canonical and personal representations to hope meaning. The diagrams should be purposeful, rather than independent artefacts, open to interpretation and open to reinterpretation. Users are able to inhabit the purposes of the diagram, in the same way that readers can be drawn into inhabiting a novel, seeing how the narrative plays out as the characters evolve: because diagrams, and elements of diagrams have connotations, just as words and paragraphs have associations through which they tell a story (Figs. 2.9 and 2.10).

Here the different representations will have different implicit and explicit connotations: as the conversation evolves, we can draw on these, hiding what we do not want. again the computer is functioning as a representation machine, encapsulating operations and encapsulating meanings. Elements of diagrams have a compressive function, just as technical words do (Sutton 1992), and these need constructing, and sometimes unpacking, to remind both teachers and children of the judgements that enable, and perhaps even constitute, that depiction.

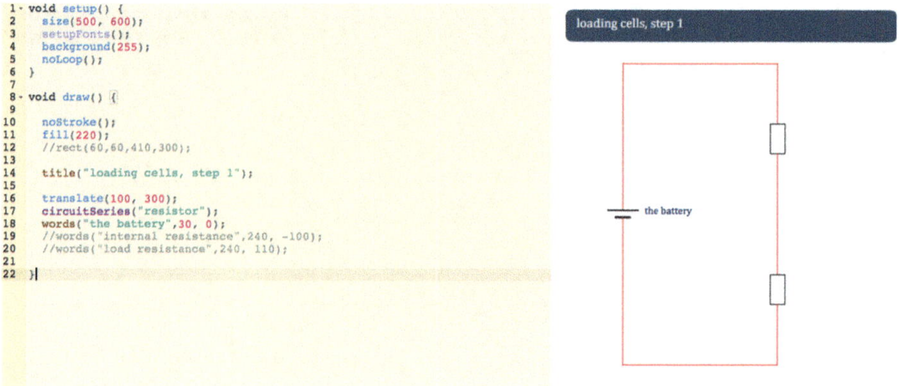

Fig. 2.6 A circuit with series connections, with the code prepared, but commented out, ready for a dialogic sequence

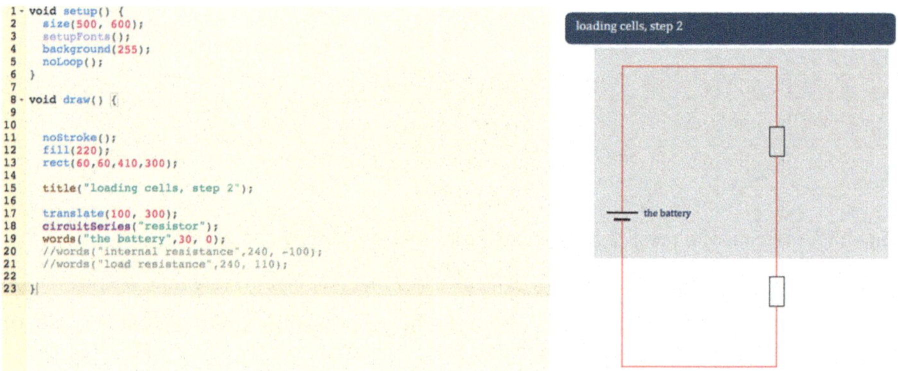

Fig. 2.7 The same circuit but now with a part of the circuit marked off as being 'internal' to be battery

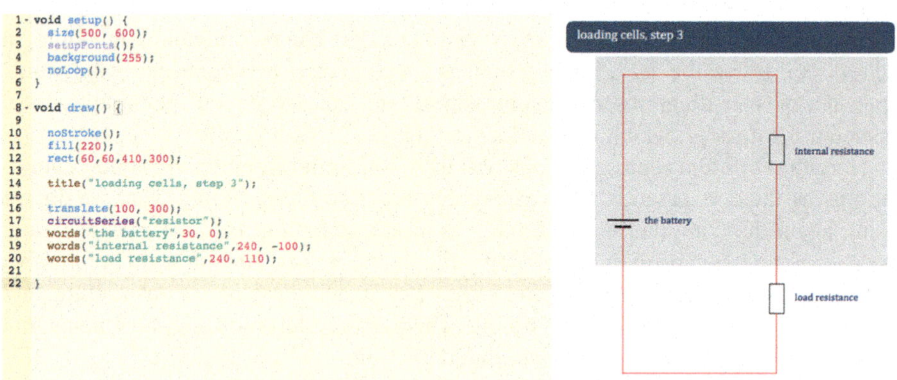

Fig. 2.8 The circuit labelled, preparatory to the next step of measuring or modelling. All of the commenting out of the code is now removed

Fig. 2.9 This is just a bulb, but it is perhaps not obvious to the learner how this very conventional representation relates to the objects that they handle in the laboratory

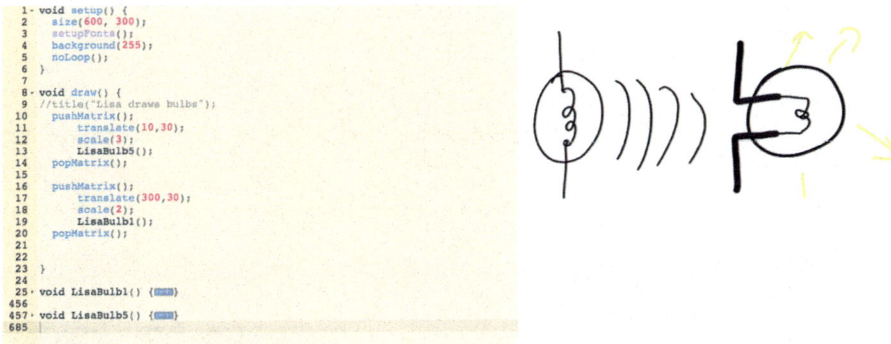

Fig. 2.10 Two versions of a diagram for a bulb drawn by Lisa, brought into the system as code and encapsulated so that they can be incorporated into more complex diagrams

Such diagrams can explore situations, exploring possibilities: one can construct and modify interactive diagrams (whether constructed and deployed as animations, simulations or models) to be used flexibly by teachers as a part of a conversation (Figs. 2.11 and 2.12).

This is, of course, simply the maximum power theorem, where we can intervene to explore the possibilities. The representations define a kind of possibility space, where the relationships we have encoded into the imagined world constrain what is possible.

It might be worth pausing to consider how this standard piece of physics is represented in other modelling systems (say Stella, Modellus, Coach) to see how these constraint relationships and the idea of a possibility space are presented. There is no time evolution here, and the chosen representation system should not nudge user into depicting the situation in ways that imply that there is.

Above all else this system should remain an open system, built to be adapted as uses evolve. There will need to be adaption at many scales, from a simple taking of one diagram and making simple alterations to rather more significant adaptions of deciding the degree of encapsulation appropriate for the conversation at hand.

```
//decisions to make
float E=12; //constant EMF of cell
//float Rinternal=3; //internal resistance of cell: constant
//load to vary, so set by dragging, value Rload
//consequences
float V; // pd across load resistor
float I; // current in loop
float P; //power dissipated in load resistor


void setup() {
     size(600, 400);
     setupFonts();
     cmr.moveTo(100,200);
     theLoad = cmr.resistance(4);
     cmr.moveTo(30,200);
     theInternal= cmr.resistance(3);


}

void draw() {
     exploreCMR();

     title("interoal resistance");

     words("internal",109, 220);
     words("load",179,220);

     Rload=theLoad.getValue();
     Rinternal=theInternal.getValue();
     I=E/(Rinternal+Rload);
     V=E-I*Rinternal;
     P=V*I;

     translate(400,200);

     power(P);
     words("power dissipated",-31, 30);

}
```

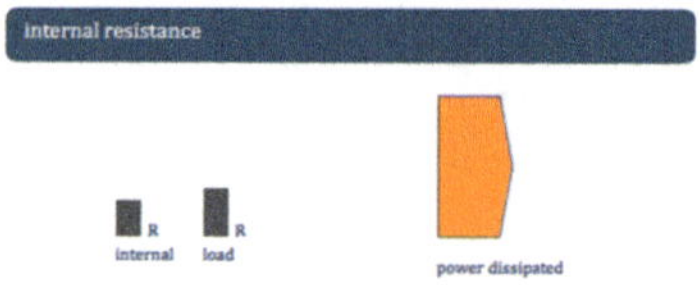

Fig. 2.11 This is a mathematical representation of the circuit in Fig. 2.8. Here you only vary the 'external' load resistance

```
//decisions to make
float E=12; //constant EMF of cell
float Rinternal=3; //internal resistance of cell: constant

//load to vary, so set by dragging, value Rload

//consequences
float V; // pd across load resistor
float I; // current in loop
float P; //power dissipated in load resistor


void setup() {
     size(600, 400);
     setupFonts();
     cmr.moveTo(100,200);
     theLoad = cmr.resistance(4);

}

void draw() {
     exploreCMR();

     title("internal resistance");

     //this should look familiar
     Rload=theLoad.getValue();
     I=E/(Rinternal+Rload);
     V=E-I*Rinternal;
     P=V*I;

     translate(400,200);

     power(P);

}
```

Fig. 2.12 Here the model is adapted so that both the internal and external resistances can be varied, so allowing the conditions under which the maximum power is dissipated to be explored

2.4 Dialogic Physics

Dialogic Approaches: Words

The dialogic approach to teaching has been advocated, and varieties of conversation have been identified to deploy in different phases of teaching. Through this, and other means, teachers have been able to become more sensitised to words and their deployment and indeed to ways in which the structure of conversations can enable the learning and structure what kinds of learning are available (Mortimor and Scott 2003).

Once Again, with Drawing

Reorganising words to make them your own may not be easy but at least has many accessible technologies to support the process. Words appear rather easy to process and both in writing and editing. Drawings are rather harder to mix and adapt.

Currently diagrams tend to be presented, rather than created or adapted, and that rather begs the question of who decides what is on offer. Treating teachers as professionals entails that some significant pedagogic decisions are made in the classroom, and this requires end-user flexibility, just as with dialogue. There may be common elements, or words, but the dialogue is not fixed. In the same way diagrams that are a part of such a dialogue accept that if developing, teachers will always adapt what you intend, so one should not make it too hard.

Do Physics: Develop Explanations

A particular concern has been to promote the doing of physics with evolving diagrams playing their part. All too often diagrams form only a part of the declarative phase of physics—representing known outcomes or certain relationships. The aim is to promote selflessness to encourage sharing the pleasure of doing physics.

Here is some physics, already done. Whilst that is a possible use of this expressive medium, it is hardly unique and not of central interest here. However it is a readable model and therefore adaptable (Fig. 2.13).

Rather I am interested in children being active participants in the representing and reasoning, in expression and exploration and in using these processes as a rich source of live metaphors for reflecting on what it is to learn physics. The work is firmly in the mental model tradition, inspired more than a little, albeit at a distance, by the work of the London Mental Models Group. The representation tools should allow range of communicative possibilities, just like language.

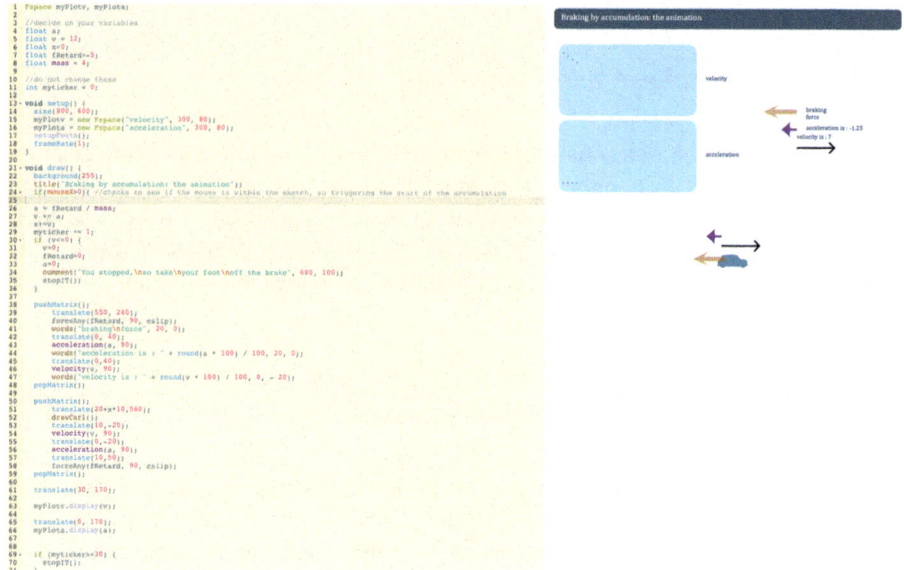

Fig. 2.13 A completed simulation: still adaptable but unlikely to be adapted by any but the most confident of teachers live in the classroom

Resonances: Coding and Modelling

I was invited to present this paper on behalf of MPTL, and it should come as no surprise that a theme is that one of the strengths of a computer as a representation machine is that it can compute: this is an ability we should exploit in thinking of them as pedagogic tools. And computational thinking itself has parallels with doing physics. Here are two aphorisms:

The art of programming is the skill of controlling complexity.
Doing physics is representing things as simply as possible, and no simpler.

Both are handy paraphrases, but there is a commonality of style that getting computational modelling going in classrooms seeks to exploit. In both cases the author has an end purpose: in the case of physics, it is to generate an imagined world that functionally mimics the lived-in world—a physics model. In the case of the programmer, the purpose is more varied, but the program will support some meaningful functionality. In this the role of the computer as a 'representation machine' is crucial. Below the surface appearances of similarities in mimicry, the core programming processes of encapsulation and abstraction provide a good basis to believe that computational models can be pedagogically useful, in describing both processes and situations in physics.

Describing and modelling situations or processes involve both abstraction and encapsulation: just considering the Earth as a point mass requires both

Fig. 2.14 Both panes, containing steps in an argument, and transitions, moving between such steps, can be incorporated into the diagrams

transpositions. This commonality of style suggests that computational modelling of situations or processes—expressing your thoughts about physics in a computational medium—could benefit learners:

- As they explore models built by others
- As they create their own models
- As they use their experiences of such explorations and expressions to reason about situations and processes

So there is something to be gained at quite a deep level by linking developments in teaching physics to the current interest in teaching children to code.

Representing a Chain of Reasoning

Dialogue often navigates a chain of reasoning, and it would be useful if the time-based evolution could be transmuted into a spatial arrangement that allows an overview of the chain. This is possible, and there may be value in having a standard graphical language for the transitions between the steps and the kinds of steps, as well as for the contents of the steps in the chain of reasoning (Fig. 2.14).

But even within a step, the readability of the code can be optimised to enable the line of argument to be followed.

It is possible that well-written code (perhaps even moving towards literate programming) may be more intelligible than any other process for connecting a chain of reasoning that relies on computation. It is often very hard to disentangle a computational model built using a spreadsheet and see the flow of the algorithm at work: indeed exactly this issue has motivated whole libraries in python to try and move data journalism on from using spreadsheets, exactly so that interested readers can check how the raw data is connected to the assertions. For the purpose at hand, the code needs to be both readable and writeable, without having to rework one's understanding of the physics or have it refracted through distorting metaphors. The chain of reasoning should be clear from the code, as far as possible.

```
void setup() {
  size(500, 400);
  setupFonts();
  cmr.moveTo(80, 200);
  batterypd=cmr.pd(4);
  cmr.moveTo(240,200);
  bulbresistance=cmr.resistance(10, label);
}

void draw() {
    exploreCMR();

    loopcurrent = batterypd.getValue()/bulbresistance.getValue();
    powerswitched = loopcurrent*batterypd.getValue();

    pushMatrix();
     translate(80, 200);
     circuitSimple("bulb");
     translate(100,0);
     power(powerswitched);
     translate(210,0);
     power(powerswitched);
    popMatrix();

    pushMatrix();
     translate(120,250);
     current(loopcurrent);
     translate(0,-100);
     current(loopcurrent);
    popMatrix();

    title("A simple circuit to explore");
    comment("drag to set the resistance of the bulb\nor the battery voltage",220, 330);
}
```

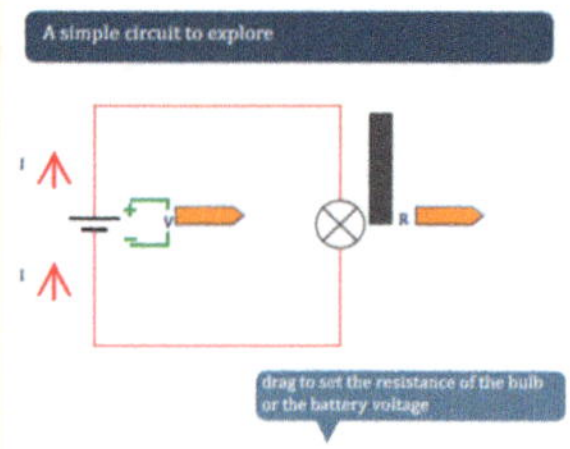

Fig. 2.15 A manipulable diagram, with the explicit code on the left. Naturally, only the diagram can be presented, whilst maintaining the interactivity

```
void setup() {
      size(650, 400);
      setupFonts();
      cmr.moveTo(80, 200);
      batterypd=cmr.pd(4);
      cmr.moveTo(240,200);
      innerbulbresistance=cmr.resistance(10, label);
      cmr.moveTo(450,200);
      outerbulbresistance=cmr.resistance(10, label);
}

void draw() {
      exploreCMR();

      innerloopcurrent = batterypd.getValue()/innerbulbresistance.getValue();
      outerloopcurrent = batterypd.getValue()/outerbulbresistance.getValue();
      innerpowerswitched = innerloopcurrent*batterypd.getValue();
      outerpowerswitched = outerloopcurrent*batterypd.getValue();
      batterypowerswitched = innerpowerswitched+outerpowerswitched;

      pushMatrix();
          translate(80, 200);
          circuitParallel("bulb");
          translate(100,0);
          power(batterypowerswitched);
          translate(210,0);
          power(innerpowerswitched);
          translate(210,0);
          power(outerpowerswitched);
      popMatrix();

      pushMatrix();
          translate(330,250);
          current(-innerloopcurrent);
          translate(210,0);
          current(-outerloopcurrent);
      popMatrix();

      title("A circuit with two loops to explore");
      comment("drag up/down to set\nthe resistance of either bulb\nor the battery voltage",320, 315);
}
```

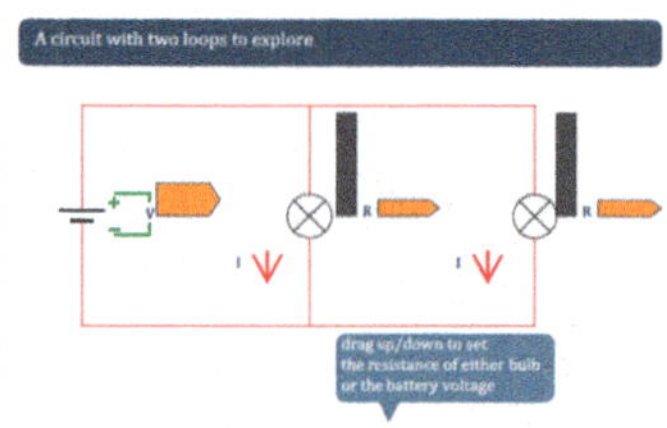

Fig. 2.16 Some simple changes in the code result in a description of a new situation, with the same flow from what is varied to the power dissipated remaining explicit

Reworking Explanations with Code

Another advantage of using code is that sequences can be traversed by reworking the code. An example is probably more powerful than general arguments (Figs. 2.15, 2.16, 2.17, and 2.18).

Doing such reworking live enables supported exemplification of physics being done, revealing it as effortful thinking (Kahneman 2011), reworking the mental modelling clay (Ogborn and Jennison 1994) until the imagined world becomes a functional mimic of the lived-in world, at least in salient and valued respects.

```
boolean twoLoop=false;

void setup() {
  size(600, 400);
  setupFonts();
    smooth();
}

void draw() {
    background(255);
    if(twoLoop==true){
     pushMatrix();
      translate(50, 200);
      circuitParallel("bulb");
      translate(55,30);
      power(6);
      translate(220,0);
      power(3);
      translate(215,-30);
      power(3);
     popMatrix();
    }

    if(!twoLoop){
     pushMatrix();
      translate(50, 200);
      circuitSimple("bulb");
      translate(55,30);
      power(3);
      translate(220,0);
      power(3);
     popMatrix();
    }

    comment("click to add or remove a loop",50, 350);
    title("Making one loop into two loops");
}

void mouseClicked() {
        twoLoop=!twoLoop;
}
```

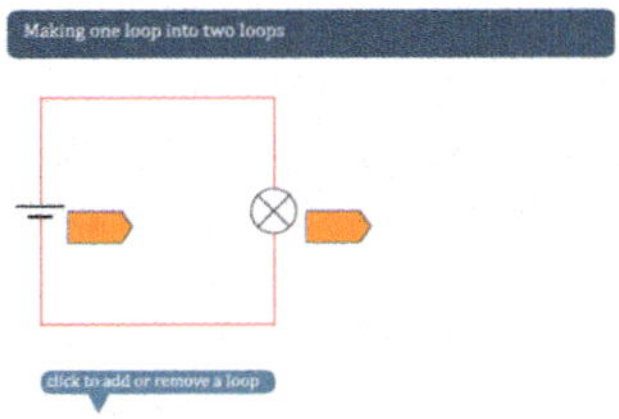

Fig. 2.17 A model to smooth the transitions between Figs. 2.15 and 2.16, where a simple click adds an extra loop

```
boolean combineLoop=false;

void setup() {
  size(800, 400);
  setupFonts();
}

void draw() {

  background(255);

    if(combineLoop==true){
     pushMatrix();
      translate(380, 200);
      scale(-1,1);
      circuitSimple("bulb");
      translate(55,0);
      power(3);
      translate(220,0);
      power(3);
     popMatrix();
     pushMatrix();
      translate(380, 200);
      circuitSimple("bulb");
      translate(55,30);
      power(3);
      translate(220,0);
      power(3);
     popMatrix();
    }

    if(!combineLoop){
     pushMatrix();
      translate(320, 200);
      scale(-1,1);
      circuitSimple("bulb");
      translate(55,0);
      power(3);
      translate(220,0);
      power(3);
     popMatrix();
     pushMatrix();
      translate(450, 200);
      circuitSimple("bulb");
      translate(55,30);
      power(3);
      translate(220,0);
      power(3);
     popMatrix();
    }

    comment("click to put the loops together\nor take them apart",50, 330);
  title("Combining and separating two complete loops");
}

void mouseClicked() {
        combineLoop=!combineLoop;
}
```

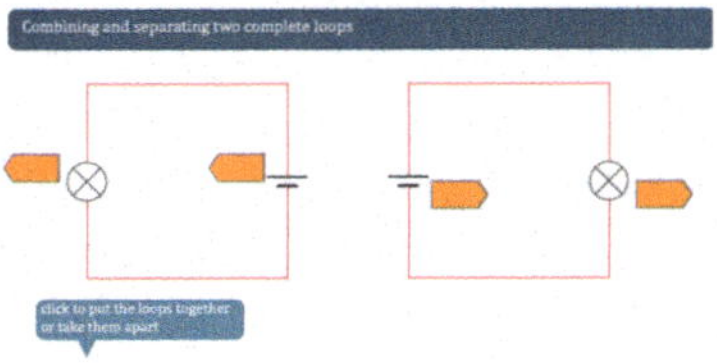

Fig. 2.18 An alternative presentation of the same move, from one loop to two, that might make the move easier to follow for some learners

2.5 Constraints in Designing a Diagram Playground

The focus here has been on reasoning with diagrams in physics and on making an amalgam of practices and tools available to teachers and children: a diagram playground. Such a playground will embody rules, perhaps as transitional objects, that facilitate access to thinking in culturally valuable ways. The possibilities for perceiving geometrical relations were fundamentally altered by the intellectual ecosystem of the turtle in Logo: a new playground was created for thinking about certain kinds of mathematical relationships and for encapsulating created patterns, which were then rather easily varied. Here I hope to make a step towards creating such a system for thinking about physics in secondary schools, as far as possible tilting the playground space so as to create a gradient that nudges teachers and pupils towards a cultured representation of physics, highlighting the fundamental epistemically moves and intellectual style.

Respecting Relationships in Physics

Physics describes both situations, where there is no time evolution of the physical quantities, and processes, where there is time evolution of the physical quantities. Both are important, but to date both have not been equally available in representational tools that are computer mediated. For example, both Stella and Modellus, which have perhaps been most widely used in computational modelling in upper secondary schools, very naturally represent time-evolving relationships such as that between velocity and acceleration but do struggle to capture the full richness of assertions of identity, such as the relationships between the quotient of force and mass and acceleration or any other constraint relationship. Given the importance of such formal relationships in elementary physics and their importance in reasoning, for example, in Piagetian explorations of compensation, such inelegance will impede the tool in supporting thinking about situations: so states. This is not simply an observation about the conflation of the assignment and equality operators. Such a slide between reasoning about time-independent states and time-evolving processes has real consequences. Viennot's work on the prevalence and perils of linear causal reasoning (Viennot 2001) stands as a continuing warning beacon as one considers the kinds of tools that one might want to make available to children and teachers: these tools should not contain a mixture of affordances and resistances that switch their thinking off down the wrong tracks.

In general, the time evolution implicit in procedural computational languages reinforces the tendency towards linear causal reasoning: creating an affordance gradient. That is, computational modelling systems exist in an Aristotelian space with a tendency to clump at their natural place: in mathematical space the corresponding location is occupied by Newtonian fluents and fluxions: formally studied as differential equations. Yet these mathematical forms and the corresponding region of space in developed physics descriptions are only really pencilled in for population later, in future studies in secondary physics schooling. Much of the action in representing and reasoning is about states or situations and not about processes or time-evolving relationships.

In designing for pedagogical utility, this affordance gradient will need to be explicitly designed out: creating structures that step outside what is naturally available in procedural computer languages. There are functional or pattern-matching languages, but the thinking underpinning these seems less well spread amongst the target audiences, and I do not find much evidence of widespread understanding and use of the style of thinking that would make them more accessible amongst teachers and children in the sciences. Indeed the extensive evidence that we have something of a tendency to explain both situations and processes to ourselves in terms of causal, time-dependent stories suggests that this absence is not at all surprising.

An expressive medium for reasoning about physics should greet both situations and processes in an even-handed way. Since the discovered rules that structure the world of physics are expressed as relationships, then such a system should be able to express constraint and time-evolving relationships equally elegantly, allowing users to express what they intend without requiring particular specialised intellectual contortions.

Both constraining and accumulating relationships should be equally prominent if the medium is to guide thinking in fruitful ways. If an approach is made through code, then some of the primitives will have to incorporate some hidden code in order to adapt any procedural language to something better suited to express constraints: accumulations can be more directly and easily expressed in procedural languages, but we may still be better building a presentational skin over the language, as is the case, for example, in Easy Java Simulations.

Implementing Both Kinds of Relationships

Here are two simple models that show how accumulations and constraints are achieved in the current implementation (Figs. 2.19, 2.20, and 2.21).

In elementary physics teaching, there are lots of situations where the descriptions require constraining relationships, but only a few where there are processes where there is a natural need for accumulating relationships. (These few are nevertheless

```
void setup() {
  size(400, 400, P2D);

  setupFonts();

  cmr.moveTo(80, 180);
  cmr.IVR(3, 6, 2, "resistance");

   cmr.moveTo(200, 350);
  cmr.aFm(3, 3, 9, "mass");

}

void draw() {

  exploreCMR();


  title("Constrained quantities");

}
```

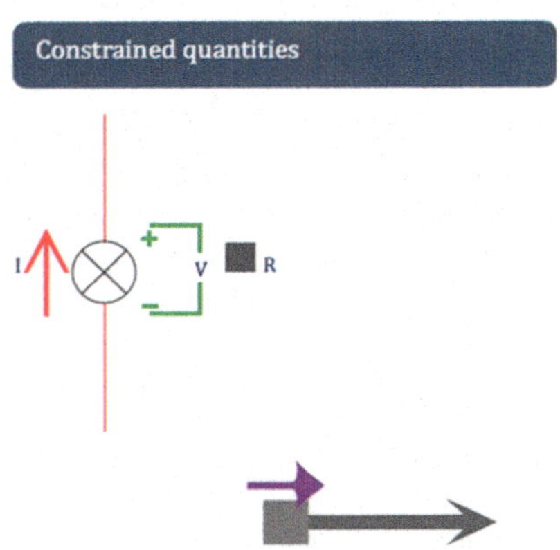

Fig. 2.19 A pair of prebuilt constraint confections. Easy to author, perhaps not so easy to read the code, which allows you to drag either the current or pd representations (the resistance is constant here) and see how varying one alters the other. You can also explore the constraints between force and acceleration in a similar manner, given that the mass is constant here

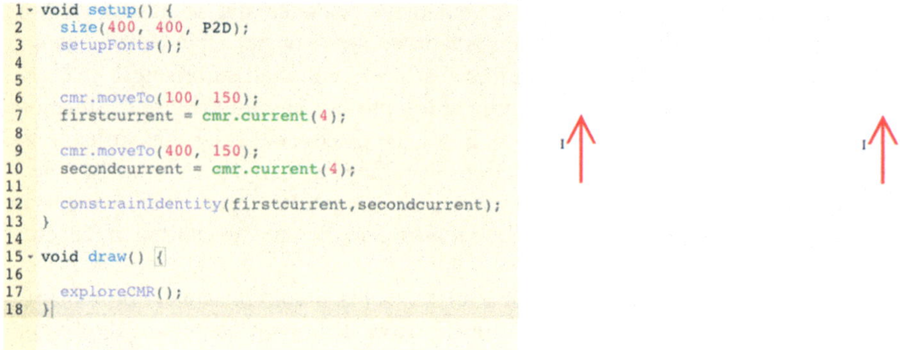

Fig. 2.20 Here the constraining is much more explicit, which may be more fruitful for the pedagogic intentions of the particular teacher

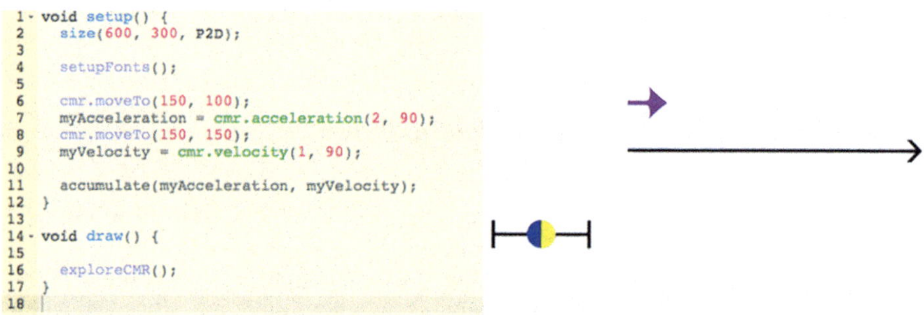

Fig. 2.21 This is an explicit accumulation relationship, where acceleration accumulates velocity. As the acceleration representation is draggable, the user is able to set both positive and negative accumulations as the model is running

important and are often core relationships, foundational for whole topic areas in physics, such as kinematics.) In existing computational modelling tools, it has been the other way around: the affordances in the tools make accumulations easy but constraints difficult—technology can derail pedagogy.

Two Paths to Developing Descriptions

Faced with a new phenomenon, an approach is to notice certain features, operationalise some identified features into a series of measurements and then hunt for patterns amongst these measures. This is a perfectly reputable analytic and empirical approach. Data capture and analysis tools (often used in MBL), prioritising creating scatter graphs as a discovery tool, lend themselves well to this function-fitting approach to the development of a descriptions containing relationships between physical quantities. Yet it is far from the only way that physicists proceed in representing and reasoning about phenomena. The theoreticians approach to representing and reasoning places a powerful focus on imagining things as they might be and seeing how that imagined world plays out. This approach is a natural fit for a computationally enabled playground for ideas as the 'playing out' of the rules is a strength of the computer: the pre-requisite is that the imagined world is simple enough to construct it and easy enough to explore. Exploration will be needed if the ability to reasoning about situations is required: it would be an unhelpful move to build in a natural assumption that the imagined world will explore some planned sequence of possibilities, however tempting that might be. Such an exploration would be much too close to an evolution and so could be mistaken for the description of a process rather than a situation. This idea of imagined worlds and the promotion of theoretical thinking are not new, reaching back to and beyond the London Mental Models group. It may actually be the purest form of a public representation of a mental model in the sense that Johnson-Laird intended, in that it is possible to see for ourselves how our world actually plays out: the computer screen may function as a non-distorting mirror for our thinking. (The computer is notorious for allowing little room for wish fulfilment in the matter of executing instructions.)

This idea of an invented world that is compared back to the lived-in world serves well as a metaphor for thinking about the whole edifice of physics as a set of provisional human constructions, there to make sense of the world. In fact a story we tell back to ourselves is but a story constrained by the need for functional mimicry. As Gaiman wrote 'We who make stories know that we tell lies for a living. But they are good lies that say true things...'. In seeking to see courteous representations of physics take hold, I see no need to play second fiddle to the writers of fiction. The case for the narrative structure of explanations has been extensively made elsewhere.

Multistep Reasoning

Some narratives, often construed as arguments leading to a particular way of seeing pieces of the world as being connected in a particular way, have a story arc that proceeds a series of connected acts. These acts, or steps, are often related in particular ways: that is, there are certain intellectual moves that we make habitually.

Being able to reveal the structure of the argument and making the moves explicit support an understanding of the thinking that is being shared and so can help supporting the development of thinking like a physicist in the children: the ability to zoom in and out. Finally, often facilitated by simple familiarity, there is the identified issue of the 'hidden moves': steps which are so natural to the teacher that they seem to need neither representing nor mentioning, but which may well blindside the children, or indeed less experienced teachers.

So a representational medium, besides expressive or metaphorical use, might also have an exploratory use: that is, children and their teachers could explore a representation or a series of linked representations, so as to unpick the thinking behind them (Fig. 2.22).

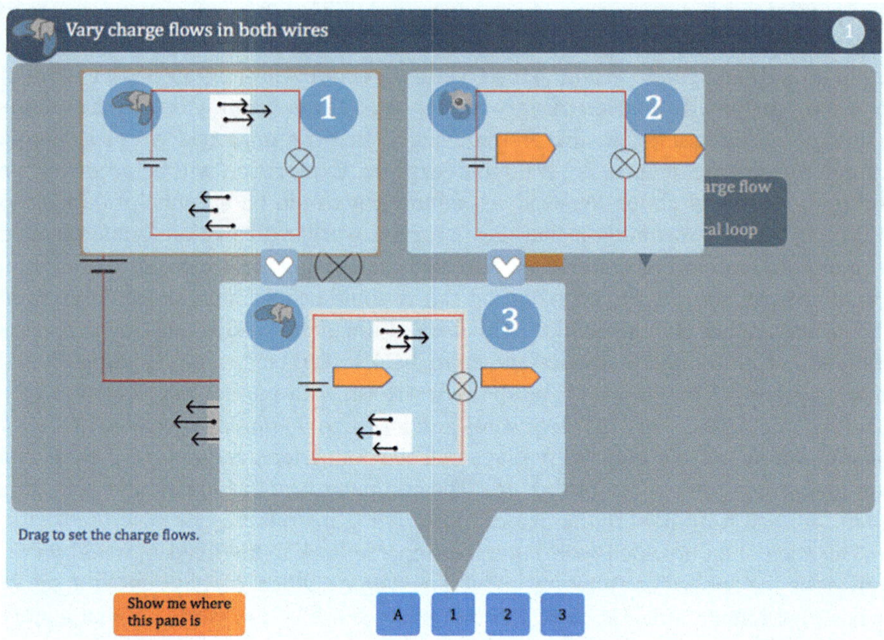

Fig. 2.22 A complex document, with an overlay showing the location of the current step in the whole three-step argument. Each step consists of an interactive diagram

This idea of switching perspective moving from one point of view to another, supported by a limited number of graphical markers, owes much to a careful reading of McCloud (1994).

Encapsulation

To think assisted by a computer, it is no longer necessary to set switches one by one—a compiler takes care of that task, allowing the user to think in more human-friendly terms. Now there are a large number of expressive computational environments, more or less abstracted from the underpinning actions of the CPU. You are able to choose a level of abstraction, selecting the representations you want to work with, so freeing up your mind to operate with the intellectual tools you choose. Of course, you can only choose from amongst available tools, and this development aims to provide one such tool from which to choose, one tuned to thinking in physics. In selecting such a tool, you are selecting a surface of thinking, able to choose to reason with entities of cultural value to physics and see how they interact in the world in which you place them. The representation to hand and the ways in which they can be induced to interact together present some affordances and resistances which could combine with practices to generate an environment where theoretical thinking in physics is encouraged.

In generating narratives there may be a need to inject representations that are not canonical and have them interact with other parts of the system. This is possible, as one can take drawn objects and render them in code, encapsulating them, as with Lisa's bulbs, met earlier.

The process is not as simple as it could be, but it's possible without huge technical overheads and could be developed further if practice supports such investment.

This move to render graphics as code might seem to introduce an unnecessary step: why not draw everything, so keeping the system entirely graphical? I think there are several good answers to that, based on current graphical programming practice and in pedagogical research.

More Natural Just to Draw?

Starting with the pedagogical, there is at least one long-term research project in getting children to create computable models based on their drawings (van Joolingen 2015). As things stand the visual representations can be drawn, with the computer understanding the drawings, using image processing. However expressing the rules that govern how these drawn entities interact is not done with drawing. I think that is because rules, expressions and intentions are not easy to derive from gestures without imposing serious constraints on the expressive range(in a way this is much like speech processing—the more constrained the environment, the better the intentions can be reliably recognised). One way of injecting these constraints is to make the drawing interface more like a vector drawing package, but use of such packages soon leads to the recognition that the presentational complexity of vector drawing software is necessary and not accidental. And such software is expensive to develop. There has been some very interesting exploratory work on dynamic drawing of the kind that might support reasoning about entities in physics, but at the moment, these are early explorations and for a more restrained range of expressive tasks that might be useful for a widely applicable tool in reasoning about physics (see, e.g. Victor 2013).

Similarly there have been construction sets based on graphical elements, sometimes to minimise syntactical errors in coding (such as Scratch) and sometimes to enable complex instrumentation to be built (such as LabVIEW). All of these graphical systems embody design decisions about what can easily be expressed in the playground, which may be more or less well adapted to the task at hand. For example, StarLogo is well suited to certain kinds of situations, but not others, and the overlap with elementary physics is not so good (e.g. in Resnick 1997).

Here we're after an open system, built to be adapted as uses evolve. For now, alphanumeric code presents the designers and users' options for graphical communication, melding structured drawings with computation, at moderate cost and with moderate commitment to possible future development paths.

A Didactical Perspective

If only all children were fluent in algebra (in expression, in manipulation and in interpretation) and as good at geometrical reasoning as Newton, then a tool such as this might be entirely superfluous. But they are not: so this diagram playground might be something of a prosthetic, to assist the learning journey. Along this journey, which is a cultured induction into physics, children should be encouraged to both represent and intervene in their collection of representations and in a manner that makes it plain to them and to others what those actions entail: that is, both the representing and the intervening should be with public representations. I am co-opting the computer as representation machine, to create a mirror for thinking in physics. That is the process of doing the physics should be as well-illuminated as possible, with exemplification of thinking in physics in the foreground.

2.6 A Step Towards a Playground: The CMR Ecosystem

For all the theorising and attempts at careful designing, the CMR ecosystem is simple and adaptable. It consists of:

- The processing.js language.
- A series of dedicated functions, written in processing.js, that allow physics representations, either interactive or static, to be shown rather simply.
- A dedicated two-pane on-line editor: in one pane a text editor to create or adapt sketches, customised to perform code completion and code highlighting with these same dedicated functions; a second pane, where the generated output appears.
- A sketch, which is a short piece of code, always contains two chunks, setup and draw, constructed using the text editor.
- A dedicated on-line player that presents the output of completed sketches, from locally saved sketch files or sketch files with a URL.
- Extensive and comprehensive documentation.

This is currently (2016) implemented and available at http://supportingphysicsteaching.net/cmr.

2.7 Moderate Progress

A Useful Tool

The aim here has been to produce a publicly accessible tool, useable by teachers in classrooms, as well as enabling preparatory work that elevates the status of diagrams in developing an understanding of physics to that of words.

In such a tool, there are a considerable number of possibilities, just as there are in a piece of apparatus, except that we're using a computer, which in its very nature is a representation machine and therefore in its essence even more flexible than a multimeter, a clock or a metre rule. But as with any piece of apparatus, it is the stories of use and practices that make it pedagogically useful and shape the future development. For now, the technology is widely used in Supporting Physics Teaching at supportingphysicsteaching.net, where it presents canonical resources that are teacher-adaptable, designed both to encourage engagement with the tool and so that teachers can adapt teaching approaches that make significant use of diagrams.

Teachers may or may not use this or similar tools: what we do know for sure is that computational modelling tools, however good, are not enough. The focus on diagrams for physics, on making those adaptable, and allowing, but not insisting on a computed model, are intended to explore a new possible approach. As is the exposure of rather simple code. Whether the implementation proves sticky and widely adopted remains to be seen: it could be made to work. However the design principles and considerations are, I think, applicable to any future tool that seeks to exploit the affordances of computed diagrams in the service of productive conversations in physics classrooms.

References

Kahneman, D (2011). Thinking, Fast and Slow. London: Penguin.

McCloud, S. (1994). Understanding Comics. London: HarperCollins.

Mortimor, E. & Scott, P. (2003). Meaning Making In Secondary Science. Buckingham: Open University Press.

Ogborn, J & Jennison, B (1994). Wonder and Delight: Essays in Science Education in honour of the life and work of Eric Rogers 1902–1990.Bristol: Institute of Physics Publishing.

Resnick, M (1997), Turtles, Termites, and Traffic Jams: Explorations in Massively Parallel Microworlds. Cambridge: MIT Press

Sutton, C. (1992). Words, science and learning. Buckingham: Open University Press

Tytler, R., Prain, V., Hubber, P., Waldrip, B. (2013). Constructing Representations to Learn in Science. Rotterdam: Sense.

van Joolingen, W (2015): "Drawing-based modelling to foster early science learning" , a plenary talk presented at MPTL 2015.

Victor, B. (2013). Drawing dynamic visualisation. http://worrydream.com/#!/DrawingDynamicVisualizationsTalk (access date 2016/12/15)

Viennot, L. (2001). *Reasoning in Physics.* Dordrecht: Kluwer Academic Publishers.

Chapter 3
Technology in Teaching Physics: Benefits, Challenges, and Solutions

Ton Ellermeijer and Trinh-Ba Tran

Abstract How to make physics education more challenging, relevant, and attractive for our high school students? How to stimulate the development of creative thinking, problem solving, and other higher cognitive skills? In many countries, governments like to stimulate science and technology in schools, and in this direction, STEM (or STEAM), MINT (Germany), and IBSE are the more recent Alphabet Soup acronyms.

Can technology applied in physics education bring us closer to the desired goals? Clearly it has been demonstrated that technology can help to make physics education more relevant, more linked to real life, and more authentic and can increase the opportunities for own investigations by the students. So it really has an added value and not just provides another way of teaching the same.

This is known for decades but still applied at a relatively small scale. A major challenge is the preparation of teachers for using technology in this direction. The authors recently investigated the development of an effective and relatively short course for teachers to prepare them for the use of ICT/Technology in IBSE lessons. The course focused not only on learning ICT skills but also on awareness of benefits and motivation. Considering these objectives, we applied several pedagogical principles, like the depth-first and one theory-practice cycle. The final course set-up is based on several rounds of try-outs and improvements and has been applied in the Netherlands, Slovak Republic, and Vietnam. The course will be presented in the full text. Some attention is given to the differences in application in different settings (pre-service and in-service, different cultures) and also the learning effects on the participants. Interesting and important conclusion is that such a high-quality course design can be applied broadly.

T. Ellermeijer (✉)
CMA, Amsterdam, The Netherlands
e-mail: ton@cma-science.nl

T.-B. Tran
Faculty of Physics, Hanoi National University of Education, Hanoi, Vietnam
e-mail: trinhtb@hnue.edu.vn

M. Pietrocola (ed.), *Upgrading Physics Education to Meet the Needs of Society*,
https://doi.org/10.1007/978-3-319-96163-7_3

3.1 Introduction

The knowledge society requires innovative products and services, and so workforce must be able to apply knowledge in creative ways. It is challenging, because many students entering technical universities are not able to cope with open tasks; they are not trained in divergent thinking. About secondary school students, there has been a decline in interest for science (OECD 2006); many school students do not see science and technology attractive, relevant, and related to jobs. Therefore, the overall challenge in physics education today is probably attracting more students; this results in driving questions: *How to make physics more challenging, relevant, and attractive for students? How to stimulate their development of creative thinking, problem solving, and other higher cognitive skills?*

In many countries, governments have been innovating the science curriculum, considering these questions. For example, to let students experience physics as a relevant topic for them, students' authentic work is included in the curriculum, and the teacher is stimulated to link regular physics lessons to real life. Special programmes to become teachers and intensive initiatives for teacher professional development are designed and implemented. To show students that physics itself is still very much alive, modern topics (e.g. quantum physics, elementary particles) and applied physics (medical imaging, biophysics) are introduced. More connections between schools and universities, research institutes, and companies are established, and more facilities for improvements of school labs are provided. To prepare students also for higher cognitive skills, the curriculum includes students' own investigations (with minds on) and design tasks (divergent thinking).

Since the 1980s, advances in technology and physics education research have stimulated intensive development of information communication technology (ICT) for data logging with sensors, video measurement, and dynamical modelling. These tools resemble those of scientists and engineers but are designed for educational purposes and primarily aimed at classroom use. Can the technology applied in physics education bring us closer to the desired goals? Clearly it has been demonstrated that technology can help to make physics education more relevant, more linked to real life, and more authentic and can increase the opportunities for own investigations by the students. So it really has an added value and not just provides another way of teaching the same. Many initiatives such as STEM or STEAM (USA), MINT (Germany), and ICT in IBSE (EC) are the more recent alphabet soup acronyms. In this chapter, we present the main contributions of technology: *showing students how physicists work today* (e.g. beats plus signal analysis, numerical model for cooling down), *enabling authentic projects by powerful tools for doing investigations* (e.g. re-entry of a capsule in atmosphere, bungee jumping, bouncing balls), and *facilitating a better link to real-life phenomena* (e.g. high jumping, parachute jump).

Educational benefits of these ICT tools are known for decades, but they are still applied in physics education at a relatively small scale. Factors involved are among others:

Limited curriculum time and limited teacher preparation time
Lack of equipment, resources, and technical support

Mismatch between assessment and curriculum objectives (e.g. inquiry)
Pupil problems (e.g. high cognitive load of inquiry learning)
Teacher problems (e.g. prescriptive instruction with ICT) and lack of continuous teacher education

Needed are systemic changes in these factors and concerted, simultaneous actions of all stakeholders. In our recent study, we focus on teacher preparation and training, which are driving forces for change in classroom practice regarding ICT incorporation. We investigated the development of an effective and relatively short course for pre- and in-service teachers to prepare them for the use of ICT in inquiry-based science education (IBSE). Several pedagogical principles, like the depth-first and one theory-practice cycle, were applied to (re)design, implement, evaluate, and optimise this ICT in IBSE course. The course was aimed not only at learning ICT skills but also at awareness of benefits and motivation. The final course set-up is based on several rounds of try-outs and improvements and has been applied in the Netherlands, Slovak Republic, and Vietnam. The course is presented in this chapter as well; some attentions are given to the differences in application in different settings (pre-service and in-service, different educational systems, and different cultures) and also the learning effects on the participants. Interesting and important conclusion is that such a high-quality course design can be applied broadly.

3.2 Benefits of Technology in Teaching Physics

An Open Computer Learning Environment

As known for a long time, ICT can enable physics education in a direction that brings students in a similar position as researchers in physics. The development of ICT tools for physics education should be driven by a combination of educational research, curriculum development, and innovative technology. We envisioned a scenario of teachers and students using a set of tools for inquiry-based study of natural and mathematical phenomena. This set of tools is integrated in one open environment designed for a broad educational setting. Openness means that it is:

- A flexible, customisable, multipurpose system
- An environment for solving open problems that need definition, set-up, exploration, data processing and analysis, mathematical modelling, and so on; that is primarily a cognitive tool
- As much as possible, free of didactic context or principles; that is less considered as a pedagogical tool but more as a tool for doing physics

This computer learning environment does not only exist in the minds of software designers, but it has already been realised to a large extent in the Coach learning and authoring environment, which is the result of three decades of sustained research and development work at CMA, Amsterdam, the Netherlands, to improve STEM education. "Coach" refers to coaching and support of learning.

A one-sentence description of Coach is as follows: Coach is a single, activity-based, open computer learning environment that is designed for the educational setting and that offers a versatile set of integrated tools for the study of Science, Technology, Engineering, and Math (STEM) (Heck et al. 2009a, b). For more detail about the latest version of Coach, please visit http://cma-science.nl/coach-7-overview.

Coach activities are mostly based on the powerful tools for *data logging with sensors* and advanced *video measurement* up to *dynamical modelling*. Teachers can use ready-made activities or author new activities to structure the lesson materials (i.e. experiments, models). Such activities may contain:

- Texts with activity explanations or instructions
- Pictures illustrating experiments, equipment, and/or context situations
- Video clips or digital images to illustrate phenomena or to use for measurement
- Measurement data presented as graphs, tables, metres, or digital values
- Models (textual, equation based, or graphical) to describe and simulate phenomena
- Programmes to control devices and to make mathematical computation links to Internet sites and other external resources for students

Particular Coach activities are presented below, but please note that all kinds of activities are supported in a single computer environment and not in a suite of separate programmes. Having a single environment instead of a bunch of special-purpose software packages brings many benefits. First, students and teachers only need to familiarise themselves with one environment, in which components are geared with each other. They can grow into their roles of skilled users of the system during their learning and teaching. A learn-once-use-often philosophy of educational tools is realisable. Additionally, students may experience the connections between different school subjects through the use of a single environment instead of a grab bag of disconnected tools. Another advantage of a single environment compared to a software suite is the possibility to easily combine different tools in one activity.

Coach Tool for Data Logging with Sensors

Characteristics The Coach tool for data logging with sensors enables students' experimentation activities in which the sensor, connected to an interface, measures a quantity (e.g. temperature, voltage, pH) in the physical world and transforms this quantity into a voltage or other signal(s), which is then read by the interface. The interface converts the signal into digital data that are transferred to, then interpreted, and processed by the connected computer or other devices with dedicated software (Fig. 3.1). A computer equipped with an interface and ample sensors becomes a universal measuring instrument, which has a wide range of sampling frequencies from very low to very high (e.g. 10,000 samples per second). This computer-based instrument certainly can take the place of instruments such as thermometers, voltmeters, and pH meters, used in conventional practical work. It enables automatic, accurate, conditional measurements and includes ample ways of storing, displaying, and analysing data. During the measurement, real-time data can be represented in graphs and tables or displayed as digital values.

3.3 Challenges of Technology in Teaching Physics

Among different software packages, Coach has common measurement methods:

- Time-based measurement in which data are automatically gathered at regular time intervals according to the sampling frequency; with this setting, it is possible to specify a signal condition from which the computer automatically starts a measurement (i.e. triggering). For example, Fig. 3.1 shows time-based measurements of interfered sound, and the obtained graph displays the sound signal.
- Event-based measurement in which measurements are taken each time a pulse (i.e. an event) is received on an interface input. For example, in an experiment with stroboscopic light, each light flash generates a pulse via a light sensor, and then a measurement is taken.
- Manually triggered measurement in which a single measurement is taken every time the user presses a button; this setting also allows to type in data (e.g. observational data, standards, given values) via the keyboard. For example, in the Boyle's law experiment, pressure of air, which is trapped inside a cylinder fitted with a piston, is measured by a pressure sensor each time a value of gas volume read by naked eyes is manually entered in the software.
- Data logging with sensors is a generic experimental tool for physics, chemistry, and biology.

Examples of Coach Data Logging Activities Figure 3.2 shows a screenshot of the measurement and signal analysis of the voice sound "eeh" recorded with the €Sense interface, which is mostly used at primary school or by beginners. The diagrams show that the sound signal is well described by a sinusoidal signal that consists of

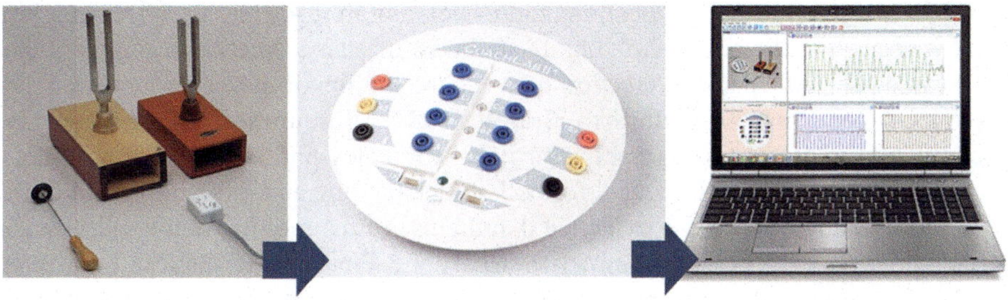

Fig. 3.1 Diagram of the tool for data logging with sensors (incl. sensor, interface, and computer with dedicated software)

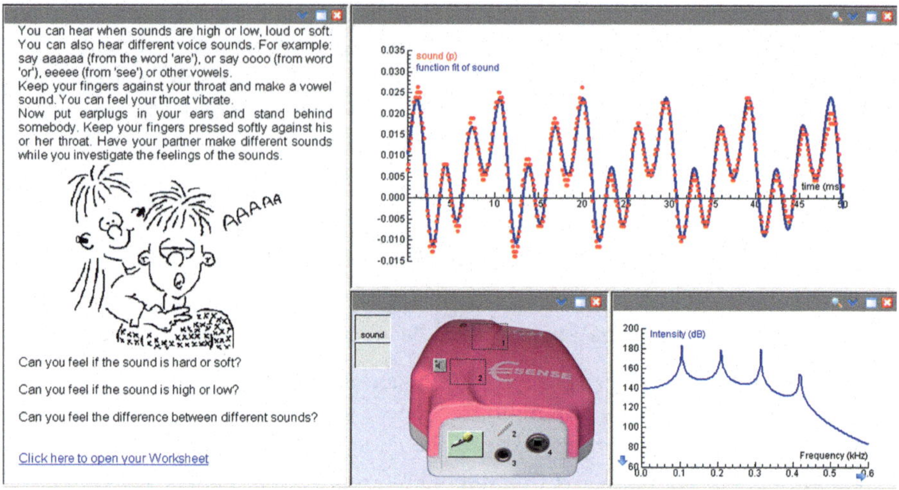

Fig. 3.2 Measurement and signal analysis of voice sounds with the €Sense interface

four frequencies. A visual representation of the €Sense interface with the built-in microphone is also present in the activity screen to make the experimental set-up clear to students. A text window is used for explanation and description of tasks.

Educational Benefits The tool for data logging with sensors has many educational benefits if it is properly integrated in the science lessons. First, this tool enhances new possibilities and contexts for physics experiments that might not be otherwise possible due to time constraints and technical difficulties (Barton 2004; Newton and Rogers 2001). This increases access to real-life phenomena, facilitates new classroom experiments, and allows measurements in the field. Second, the tool enables collecting, recording, and representing of many data and even repeating this process several times in short time (physical world). Consequently, pupils will have time in the classroom to design the experiment, interpret data, and/or explain relationships (theoretical world).

Third, the "real-time graphing" feature of the data-logging tool stimulates pupils to move back and forth between the physical world and the theoretical world. For example, a pupil walks in front of a motion sensor, and immediately the software shows in the graph her or his position and/or velocity in real time. By observing the pupil walking and the graph showing up at almost the same time, other pupils in the class can easily realise the connection between the motion of their classmate and the kinematics concepts. According to Brasell (1987), this immediacy between the phenomenon and real-time graphing of data stimulates pupils' conceptual understanding, and this feature is critical for both understanding and motivation. Sokoloff et al. (2007) showed research evidence that the use of the tool for data logging with sensors in a laboratory curriculum (i.e. RealTime Physics Mechanics) improved pupils' understanding of dynamics concepts, and the retention of the concepts by those pupils was excellent.

Last but not least, the incorporation of the data-logging tool enables pupils to participate in aspects of scientists' experimental inquiry, considering that the data-logging tool is similar to those used by scientists. According to Ellermeijer et al. (1996), once pupils get used to the data-logging tool, they can decide and reflect at any time about what to measure, how to calibrate, and what readings should be taken. This shows that such participation in authentic inquiry with the data-logging tool will stimulate pupils to comprehend scientific inquiry.

Coach Tool for Video Measurement

Characteristics The Coach tool for video measurement enables pupils to conduct experiments in which, for instance, position and time data of a moving object, registered in a digital video, are collected in the successive video frames by mouse clicking. Among different software, Coach includes common steps to gather real-life data from a video. First, the user has to define the video scale, time calibration, and coordinate system. The video clip is scaled by specifying which distance on the video screen corresponds to which actual distance (e.g. 1 m viewed in the video frame in Fig. 3.3). A video is a collection of rapidly displayed pictures called video frames. The time interval between two successive frames shown in the software is calibrated by entering the actual frame rate of the video (i.e. how many video frames were taken in a second as the video was recorded). Next, the user moves the cursor over the video screen to locate the point(s) of interest (e.g. a baseball) and then click to store the first video point (i.e. first coordinate and time data). The video clip automatically advances to the next frame, and then the user continues clicking on the

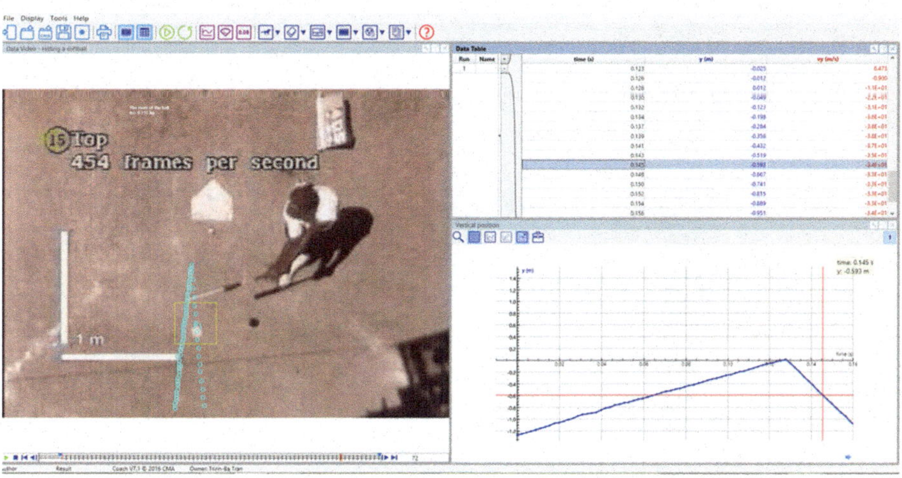

Fig. 3.3 Screenshot of an experimentation activity facilitated by video-measurement tool

reference point. This procedure with the software is repeated until the user obtains a desired number of data points.

Coach allows automated tracking of the movement of objects and enables collection of different video points in a single video frame. Like the data logging with sensors, during the manual measurement and automated tracking from a video, the collected data are simultaneously displayed in a diagram or table (real-time graphing) (Fig. 3.3). Other dynamics quantities such as velocity, acceleration, momentum, kinetic energy, and force can be numerically computed based on the collected data. Finally, collected and computed data are analysed and processed further by the software. Video measurement is mostly limited to movement, so it is mostly used in physics.

Examples of Coach Video Measurement Activities Figure 3.3 illustrates an experimentation activity facilitated by the video-measurement tool. In this activity, position and time of a baseball are collected from a high-speed video and displayed in the graph and table by the software. The dotted cross in the graph indicates that the scan feature of the software is activated. In this illustration, the data point (−0.593 m, 0.145 s) on the graph and the table is scanned, and the video advances to Frame 72, which shows the corresponding position of the baseball.

Another example is the video measurement of the motion of a self-made yoyo, which is winding up and down (Fig. 3.4). In this case, the position of the point near the rim of the disk and marked by a sticker (P1) is measured in a slightly moving coordinate frame whose origin is at the hand of the person holding the end of the cord of the yoyo. But point tracking makes the measurement at hand easier and less time-consuming: In the starting frame, the positions of the hand and of the sticker are specified, and the shapes of the search areas (white boxes) are set, and then the coordinates of these points are automatically recorded in subsequent frames.

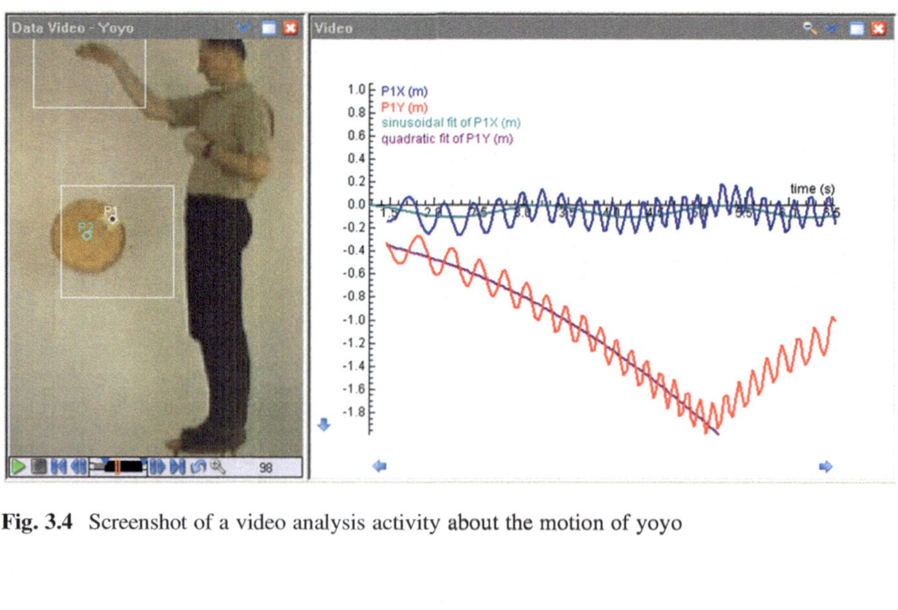

Fig. 3.4 Screenshot of a video analysis activity about the motion of yoyo

In the diagram to the right, the horizontal position and the vertical position of P1 are plotted against time. This is combined in the diagram with a sinusoidal fit of the horizontal displacement of the yoyo, due to an unintentional pendulum motion of the yoyo, and a quadratic function fit of the vertical position during the first phase in which the yoyo unwinds. These trend curves can be used as coordinate functions of a computed point that is displayed in the video clip (P2): It turns out to be close to the position of the axle during the unwinding phase of the yoyo. Please refer to Heck and Uylings (2005) for detailed modelling of the yoyo motion.

Educational Benefits The Coach tool for video measurement has much added value if it is incorporated appropriately in school science. First, like the data-logging tool, video measurement creates new possibilities and contexts for experimentation activities. With the video-measurement tool, the teacher can bring real-life, attractive scenes of motion into classroom activities that show pupils the relevance of science concepts and theory in everyday life (Heck 2009; Zollman and Fuller 1994). Such realistic scenes of motion can be quite ordinary (e.g. basketball shots, amusement-park rides, dancing) or unusual (e.g. car crashes, jumps on the Moon, rocket launch). With high-speed videos (i.e. up to 1200 frames per second), the teacher and pupils can quantitatively explore many more situations of realistic motions (e.g. multidimensional collisions between billiard balls, gun recoil) that would be mostly impossible to investigate with traditional instruments and even with sensors for school science. Additionally, the video-measurement tool can serve as a cost- and time-effective instrument for the school laboratory, which might replace rulers, timers, photogates, and motion sensors in motion-related experiments.

Second, the tool enables the collection and representation of many video data from different realistic situations in a short time (physical world). Consequently, pupils will have time in the classroom to interpret data and/or explain relationships (theoretical world). Third, the "real-time graphing" and "scan" features of the video-measurement tool stimulate pupils to think back and forth between the physical and theoretical worlds. This becomes more likely as images of these two worlds are shown in the same software interface (Fig. 3.3). When pupils scan a particular data point in one of the graphs, the corresponding video frame, where the data were collected, displays simultaneously. This feature enables pupils to identify events during the realistic situation (physical world) and connect them to abstract representations in the graph (theoretical world). This results in pupils' deeper understanding of the motion and related kinematic concepts (Beichner 1996; Gröber et al. 2014).

Last but not least, the incorporation of the video-measurement tool makes it possible for pupils to exercise experimental inquiry practices similar to those of biomechanics and movement-science scientists (Heck 2009; Kearney and Treagust 2001; Laws and Pfister 1998). Pupils can participate in many aspects of experimental inquiry using video measurement, for example, formulating problems; designing the scenario and set-up for appropriate video recording by a webcam, a smartphone, or a video camera; calibrating time and scale of the video; defining from which frames to get data and with which techniques to collect data; and processing and interpreting the collected video data.

Coach Tool for Dynamical Modelling

Characteristics Modelling has different meanings for different communities, depending upon the context in which it is discussed. The term "modelling" will refer to computational, dynamical modelling that is a tool used by scientists in many different fields (e.g. science, technology, economics, sociology) to describe, explain, and predict complex dynamical systems. It helps to understand a system's structure, the interaction between its objects, and the behaviour it can produce. Many of such systems can be built as models on the computer, which can carry out many more simultaneous calculations than human mental models and which can enable solution of differential equations. These differential equations cannot be solved with secondary school mathematics.

The Coach tool for dynamical modelling provides the teacher and pupils with possibilities to be engaged in the modelling process in science: "analyse a situation in a realistic context and reduce it to a manageable problem, translate this into a model, generate outcomes, interpret these outcomes, and test and evaluate the model" (van Buuren et al. 2010, p. 112). First, a realistic context (e.g. a tennis ball bouncing on the floor) is analysed and simplified to be manageable by ignoring realistic effects or situational factors (e.g. the ball moving vertically without rotation, air resistance, and aerodynamics effects); the stripped-down, mental model is then translated into a computational model. Next, the computational model is constructed by graphical elements: state variables (e.g. height, velocity); in- and outflows of state variables (i.e. rates of change); auxiliary variables; constants (e.g. acceleration due to gravity); events (e.g. bounce) that provoke discrete, instantaneous changes of state variables; and relations that are visualised by connectors between variables, constants, and events (Fig. 3.5) and are specified by simple mathematical formulas.

As the model is executed, differential equations behind the model are automatically solved by numerical iteration methods and so result in values of variables as a function of time. To interpret these modelling data, the modeller needs to choose relevant representations of the resulting values of variables such as (a) graphs that show more explicit, comprehensible relationship between variables and (b) animations that visualise behaviours of modelled objects. To validate the model (i.e. evaluating its descriptive, predictive, and explanatory quality), the modeller compares modelling outcomes with their counterparts in the physical world (i.e. standards, measured data, empirical graphs).

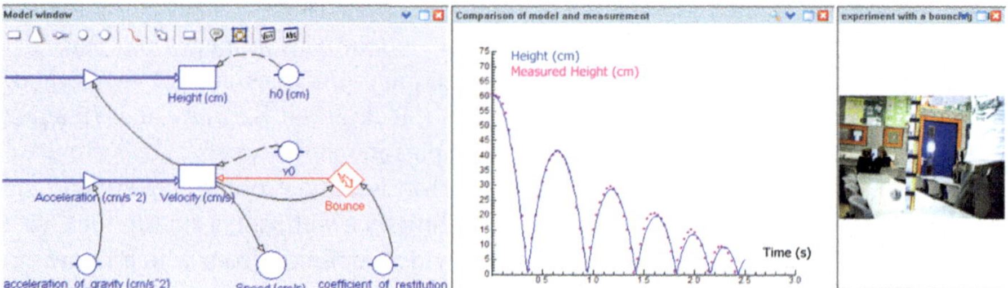

Fig. 3.5 Screenshot of a modelling activity facilitated by the modelling tool

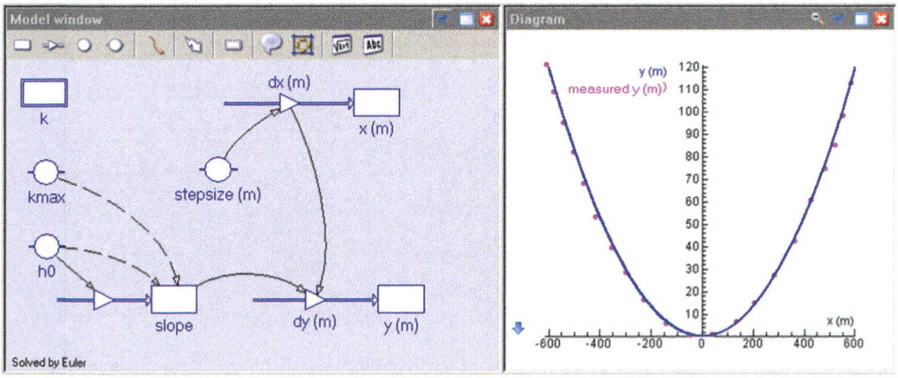

Fig. 3.6 A graphical model of the shape of the Golden Gate Bridge

There are different ways to represent variables and relationships behind a dynamical model in Coach, including (a) the stock and flow mode (graphical representation) and (b) text-based modes, using equations or a textual representation. In this chapter, we confined ourselves to graphical modelling with a stock and flow representation of variables and relationships among these variables. This was because the stock and flow representations stimulate pupils to focus on qualitative relationships (theoretical world) and connections of these to the realistic situation (physical world) rather than mathematical equations or programming syntaxes.

Examples of Coach Modelling Activities Figure 3.5 illustrates a modelling activity facilitated by the modelling tool. In this activity, bouncing of a solid, rubber ball is modelled, and the modelling result is compared with data obtained from video measurement of the bouncing ball. The graph shows the modelling result (solid curve) and the measurement (dots) for height versus time. Another example is a graphical model of the main span of the Golden Gate Bridge (Fig. 3.6), which is based on the approximation of the suspension cable by k_{max} straight line segments with horizontally equidistant joint.

Coach is in fact a hybrid system that combines a traditional system dynamics approach with event-based modelling. The left window of Fig. 3.7 shows a graphical model of a ball hanging on a vertical spring attached to the ceiling and that can also bounce against the ceiling; a special event icon (with the thunderbolt symbol) is used to specify what should happen when the ball bounces. The window in the middle is an animation window that displays the simulation results as animations where model variables are presented as animated graphics objects. A student can interact with the animation through a slider bar, that is, select the value of the spring coefficient before the start of the simulation or change it while the simulation runs. Animation allows students to first concentrate on understanding a phenomenon with the help of simulations before going into the details of how the simulations have been implemented by means of computer models.

Educational Benefits First, the modelling tool holds the potential to enlarge possibilities for pupils' theoretical inquiry of realistic, dynamic phenomena (e.g. motion

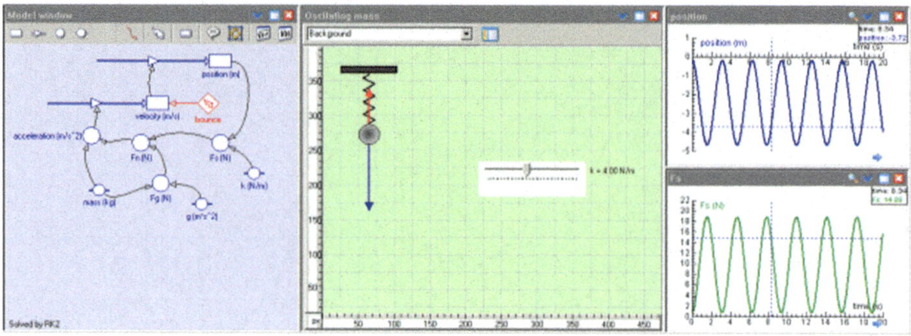

Fig. 3.7 A model of the harmonic motion of an oscillating ball hanging on a spring and an animation of the motion

with air resistance, charging and discharging capacitors, combustion of carbon monoxide, and chemical equilibrium). These phenomena are difficult to describe with school mathematics but relatively easy to model with software (Heck 2009; Velanova et al. 2014). There are different patterns in which pupils can move back and forth between the theoretical and physical worlds and so learn with the modelling tool. For example, pupils run a given model (e.g. a parachute jump with air resistance) to understand a phenomenon and/or explore its structure to gain insight into interactions between the model elements. Based upon their understanding, pupils can also make a small change to a given model, try out various modelling ideas, and then evaluate if the revised model describes the phenomenon better. For example, "unfortunately, the parachute does not open right away. Therefore, there is first free-fall for two minutes and then fall with air resistance while the parachute already opens". With a certain mastery of the modelling tool, pupils may construct a new model from their mental model of the realistic phenomenon and validate the model by comparing modelling outcomes with experimental results. Patterns for teachers to prepare a lesson, using the modelling tool, are similar; the teacher might use a ready model, modify it a bit, or develop a new model.

Second, the software allows importing measured data and graphs to the modelling activity. This enables simultaneous observations of the modelling graph (i.e. an outcome from the theoretical world) and the experimental graph (i.e. an outcome from the physical world) in the same diagram (Fig. 3.5). It is convenient for pupils to compare these outcomes of the two worlds. If the modelling result does not fit the real data, then pupils can adjust the model (e.g. changing parameters, adding variables, correcting relationships), execute it again, and compare new modelling results with the real data. The modelling tool enhances opportunities for many rounds of thinking back and forth between the theoretical and physical worlds.

Last but not least, the incorporation of the modelling tool enables pupils to (a) get used to modelling as a scientific tool in computational science (doing science with computer), (b) appreciate what modelling is as a way of thinking, (c) understand how important it is in science, and (d) develop a critical attitude by working with several models for one and the same phenomenon (i.e. modelling cycle). In their article, submitted in November 2008, Heck and Ellermeijer (2009) used Coach video measurements and models of runners to predict the possible time: 9.6 s for 100 m of Usain Bolt based on his Olympic run in Beijing 2008. This model accurately predicted or apparently affirmed his world record at the 2009 World Championships in Athletics in Berlin (9.58 s). This instance illustrates the power of the Coach tools in explaining and predicting real-world phenomena like sprinter's run. Additionally, Heck and colleagues showcased students' research projects (i.e. yoyos, alcohol metabolism, beer foam, bouncing balls) in which pupils could build models from simple to more complex (i.e. progressive modelling approach) by incorporating more factors aimed at better matching between the model and reality (Heck 2007, 2009; Heck et al. 2009a).

Data Processing and Analysis in Coach: Generic Components of the Coach Tools

Generated from the model or collected from the experiment or the video, numerical data can be then quickly transformed into more comprehensible, graphical representations: graphs, tables, and animations. If such representation forms are arranged in advance, then pupils can see, for example, how empirical graphs appear (i.e. real-time graphing) or how animated objects move during the measurement or generation of data.

Additionally, just requiring simple manipulations, the software provides pupils with many possibilities for elementary analysis such as scan, slope, and area and further processing such as (a) fitting or modelling the data with analytic functions (e.g. function fit), (b) integrating and differentiating data, and (c) displaying Fourier transforms of the data (Sokoloff et al. 2007; Heck et al. 2009b). Moreover, swift analysis and processing with three tools save time on labour-intensive, repetitive tasks (e.g. drawing graphs) and so allow pupils to focus on inquiry skills like interpreting data, inferring relationships, and testing different assumptions, which are otherwise impossible due to time constraints. These features of data processing and analysis crucially add to specific characteristics of the three Coach tools mentioned above. This makes each of the ICT tools an authentic platform where pupils can easily move back and forth between the physical and theoretical worlds within the classroom time to generate or validate knowledge of science.

An Example of ICT in a Student Project: A Surprising Result

The Student Project with ICT: "Physics of Bungee Jumping" The Dutch curricula require pupils to gain exposure to research projects in physics and other subjects where they have to make their own choices with respect to topic, questions, and experiments/models; collect and analyse data; and compare outcomes with literature. A final investigation project is intended for 80 h outside of regular lessons and spread over a whole school year (about 2 h a week). In 2003, two Dutch students teamed up

Fig. 3.8 Graphical display of experimental results (below) and computed values (above)

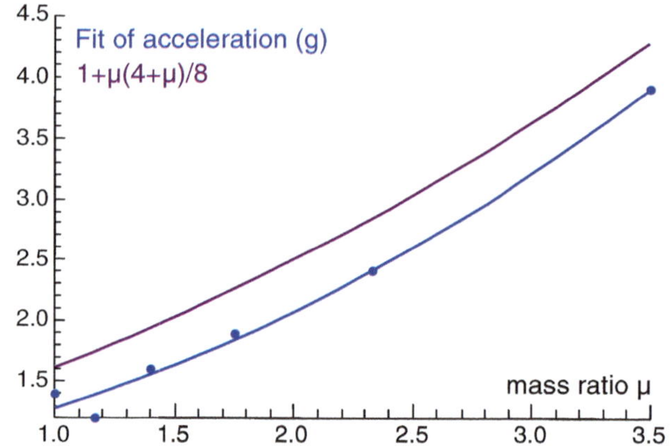

to investigate the physics of bungee jumping. In the first phase of bungee jumping, the bungee jumper falls down, and the bungee rope is still slack. In instructional material, this phase is often considered a free-fall. Taking into account the mass of the bungee rope, the students formulated the research question: "*How large is the acceleration at a bungee jump and to what degree is this acceleration influenced by the relative mass of the rope and the jumper?*"

The students collected position-time data through Coach video measurements on a dropped scale model (an Action Man toy figure) and on dropped wooden blocks of various weights attached to ropes of various stiffness. The velocity and acceleration of the dropped object were computed by numerical differentiation. Soon the students realised that the mass ratio between rope and objects was too low to see an outstanding result and they repeated the experiment with objects of larger mass ratio. The graph of the acceleration at the moment that the block has fallen a distance equal to the rest length of the elastic as a function of the mass ratio of elastic and block is shown in Fig. 3.8, together with the graph of the following theoretical result:

$$a = g\left(1 + \frac{\mu(4+\mu)}{8}\right) \tag{3.1}$$

where μ is the mass ratio of the elastic and the wooden block. The students noted that the graphs obtained by measurement and theory are alike, with the theoretical values just a bit higher. They attributed the difference mainly to the development of heat during the motion.

The Surprising Result Not knowing that a Dutch physics teacher had published around the same time about an experimental verification of the physics of bungee jumping, the students wrote an article about their work that was published in the journal of the Dutch Physics Society. The students' article claimed the result for acceleration during first phase "free-fall" up to $a = 3.9$ g for mass ratio m/M = 3.5. It triggered quite a number of reactions in the journal and for almost a year on the Internet. It seemed that a major part of the physics community, at all levels of education, was suddenly playing with ropes, chains, elastics, and so on. The result of the student project is contrary to the usual experience with free-falling objects and therefore hard to believe by many a person, even by an experienced physicist.

It was a starting point for heated discussions about the quality of the experiments and the physics knowledge of the experimentalist, and it even prompted complaints about the quality of current physics education in the Netherlands. However, experiments did reveal the truth, and students could do this supported by ICT tools. Two theoretical physicists agreed with the findings of the students, and they explained that physics intuition is easily fooled, as everyone is taught the Galilean paradigm of the motion of constant masses, according to which acceleration must be produced by a force. A launched rocket and a falling chain or slinky are important counterexamples to this line of thought. Actually, as can be seen in the theoretical section, believing the statement $a > g$ means giving up or generalising the law $F = ma$. For other bungee-jumping experiments, which investigate the phenomenon further and make more use of the Coach, we refer to Heck et al. (2010).

Integration of the ICT Tools in Recent Physics Curricula

Already for a long time, physics curricula have included laboratory activities. Together with a widespread integration of ICT in schools, there have been more and more school curricula that incorporate data logging with sensors in the science laboratory. Furthermore, in recent years, learning about modelling and learning to model have become explicit goals of science curricula in many countries (e.g. the Netherlands, the United States, Germany, and the United Kingdom). The current framework for science education in the United States (NRC 2012), for example, stated that:

Curricula will need to stress the role of models explicitly and provide students with modelling tools so that students come to value this core practice and develop a level of facility in constructing and applying appropriate models. (p. 59)

Modelling is now included in the Dutch physics curriculum (and also in chemistry and biology) for the preuniversity track in secondary schools. Recently, a learning path to achieve modelling skills has been developed for the Dutch lower secondary physics curriculum. This learning path is completely integrated into the curriculum and has been tested in school practice (van Buuren 2014).

Looking back at 30 years of research on the use of ICT in education, curriculum development, and software/hardware development, it is fair to say that a lot has been achieved. Hardware and software development, including the development of the working environment—Coach, has been able up to now to meet more or less the requirements of trends in STEM education such as the change towards context-rich education, emphasis on scientific approaches, better preparation for higher education through a stronger focus on competencies, and emphasis on individual learning and provision of students' autonomy over the process of knowledge and skills acquisition. This work will undoubtedly continue, due to the very nature of technology and education and new demands from society.

Conclusions About Added Values of ICT in Physics Teaching

It has been demonstrated that the Coach tools for data logging, video measurement, and modelling can stimulate inquiry by pupils. The proper use of these tools enhances opportunities and time for pupils' generation and validation of knowledge in the classroom. These meaningful opportunities and sufficient time enable pupils to move back and forth between the physical and theoretical world within the inquiry process. Students can work directly with high-quality, real-time data in much the same way professionals do. Physics learning with the ICT tools resembles practice in contact with current research work. With these ICT tools, investigations are characterised as being challenging, complex, open-ended, and cross-disciplinary and requiring strong commitment and broad range of skills.

3.4 Challenges of Technology in Inquiry-Based Teaching of Physics

Integration of ICT into IBSE in Teaching Practice

Inquiry-Based Science Education (IBSE) Science educators have been aware of the potential benefits of an inquiry-based approach in science teaching and learning at both primary and secondary levels, and the term "inquiry-based science education" (IBSE) has been popular for a long time. In an article published in 1910, Dewey remarked that science is not only a body of knowledge to be acquired, but it also includes inquiry methodologies to generate and validate knowledge. In this chapter, we consider inquiry as a process of generating and validating knowledge through moving back and forth between the theoretical world (ideas, concepts, relationships, theories, and models) and the physical world (objects, phenomena, observations, measurements, and experiments). According to Van den Berg (2013), ideally, IBSE will engage pupils in thinking back and forth between these two worlds like scientists; and "the phenomena and experiments serve as a source for validating ideas and theories and as a playground for generating new ideas and theories in a complex mix of inductive and deductive mind play" (p. 75).

Inquiry as process of generating and validating knowledge fits into a view on learning as knowledge creation, discussed by Paavola et al. (2004). Inquiry, under the knowledge creation perspective, is the process whereby new knowledge and understandings are (re)constructed. From the knowledge creation perspective, knowledge is not always objectively true. Knowledge is not always given by teachers and scientists or in other knowledge containers (e.g. journal articles, textbooks). Knowledge and its representations (e.g. ideas, concepts, relationships, theories, and models) can also be created, elaborated, and restructured by learners and researchers. This is in line with Duschl et al. (2007) that the brain is filled with preconceptions from early-life experiences; some of these preconceptions match with science, others do not. Therefore, much learning involves reconstruction of prior ideas, which are already in the learner's brain. In addition to the knowledge creation model, Paavola et al. (2004) discussed two other metaphors of learning: acquisition and participation. The knowledge-acquisition metaphor focuses on learning within individuals' minds, whereas the participation metaphor emphasises learning as a process of participation in various practices and activities. The knowledge creation perspective encompasses both acquisition and participation.

In the book *The scientist in the crib*, Gopnik et al. (1999) implied that from young ages, children can create new knowledge by inquiry, and scientists make the most of this capacity, which lets "children learn so much so quickly" (p. 9). Consequently, we indeed concur with Duschl et al. (2007, p. 83) that pupils are able to "engage in and profit from instruction that incorporates relatively complex scientific practices from the very beginning of their schooling".

The science education community has suggested making authentic inquiry of science more accessible to pupils (e.g. Gaskell 1992; Edelson 1998; Braund and Reiss 2006). Authenticity of inquiry in the school can be interpreted as resemblance of pupil activities to experimentation/modelling activities of practicing scientists in constructing new knowledge, considering the three following aspects (Heck 2009):

- A real-life context for learning that provides pupils with opportunities to investigate realistic science problems in history or present-day research and so pupils will appreciate the relevance of scientific knowledge in everyday life.
- Tools and techniques that enable pupils to carry out experiments/modelling and to analyse and process high-quality data in much the same way scientists do.
- Scientific attitudes of learning that stimulate pupils' pursuit of unanswered questions, commitment to challenging tasks, and social interactions (e.g. cooperation, argumentation).

Authentic inquiry is close to real science and so makes school science more attractive and relevant. Moreover, considering the "learning as participation" metaphor (Paavola et al. 2004), pupils can appreciate inquiry as a scientific method of generating and validating knowledge through being engaged in practices similar to those of scientists.

IBSE is an integral component of the intended science curricula in many countries (Jeskova et al. 2015). In such school science curricula, "inquiry" refers to learning goals (i.e. understanding of the methods of scientific inquiry and the ability to carry out scientific inquiry). "Inquiry" also refers to teaching strategies that stimulate and support pupils to exercise inquiry practices, including hands-on activities, minds-on discussions, and meaning making (Hodson 2009; Minner et al. 2009; NRC 2012).

ICT and IBSE in Teaching Practice ICT, IBSE, and ICT in IBSE are still not implemented sufficiently and properly in the classrooms in most countries. Abrahams and Millar (2008) summarised results of observations in 25 typical laboratory lessons in the United Kingdom as follows:

> Practical work was generally effective in getting students to do what is intended with physical objects, but much less effective in getting them to use the intended scientific ideas to guide their actions and reflect upon the data they collect. (p. 1945)

Teachers sometimes try to apply ICT tools (e.g. data logging with sensors) in the classroom but mostly in traditional ways in which they provide a prescriptive list of tasks for pupils to follow ritualistically. Meanwhile, pupils know in advance from the textbook what the results of the practical work should be (Hofstein and Lunetta 2004). Prescriptive instruction enables pupils to operate the ICT tool on their own (manipulation of equipment), but such instruction limits inquiry opportunities for pupils (manipulation of ideas).

International ICT projects (e.g. KLiC and ICT for IST) highlighted possibilities and good practices of ICT tools for IBSE but were not of the scale needed to bring about the substantial impact on classroom implementation. At the national level, in many countries (e.g. Greece, Russia, and Ireland), governmental education-reform projects have invested in ICT apparatus and software, delivered it to schools nationwide, and provided teachers with short training. Meanwhile, knowledge-oriented curriculum objectives and paper-and-pencil assessments were not reformed. Eventually, sufficient and proper implementation of the ICT tools was not realised. In this situation, projects with a huge funding failed to make the intended change in the classroom due to lack of consistent and concerted efforts to design, align, and implement curriculum innovation regarding ICT integration.

Integration of ICT in IBSE: Challenges to Pupils

Pupils' authentic inquiry in the classroom should be similar to but cannot be the same as those of scientists, because interests, background knowledge, and motivations of pupils are enormously different from those of scientists (Edelson 1998). "Real" authentic inquiry is open-ended and requires a solid discipline-knowledge base. As Ogborn (2014) claimed, real inquiry takes some years and requires scientists' "full critical attention", whereas "replicated" inquiry in the classroom is often intended for very limited time (i.e. half an hour) and relies on pupils' "intuitive responses". These responses "will most often be wrong or misguided, yet seem good to them, and be difficult to counter" (p. 42). Reviewing a number of studies that empirically examined the inquiry learning process, de Jong and van Joolingen (1998) identified "intrinsic problems" of pupils in inquiry learning. These problems are related to hypothesis generation, design of investigations, interpretation of data, and regulation of learning. Consequently, IBSE might generate a heavy cognitive load for pupils (Kirschner et al. 2006).

Integration of ICT in IBSE: Challenges to Teachers

Even without ICT integration, proper implementation of IBSE activities is still a problem for teachers. A number of studies have reported that teachers often find it difficult to elicit pupils' ideas about the research questions; to guide pupils in planning, executing investigations and analysing, and interpreting data; and to manage pupils' independent learning at different paces (Davis et al. 2006; Hofstein and Lunetta 2004). ICT provides innovative tools for IBSE, but the use of these tools will "further complicate the complex web of overlapping factors, which characterise pedagogical thinking involved in planning and executing lessons" (Rogers and Twidle 2013, p. 229).

Effective ICT integration assumes that teachers have to learn possibilities of the ICT tools for their subject, acquire skills to operate the software and hardware, and get used to troubleshooting technical problems. More importantly, teachers need to adapt and improve their pedagogical knowledge to be able to design suitable ICT in IBSE activities and engage pupils in implementation of such activities in the classroom.

Inquiry Teaching Versus Prescriptive Instruction with ICT Pupils need sufficient instruction and practice time in order to handle laboratory equipment in the classroom. Recipes like "do this, then do that" might be the fastest and most convenient way in helping pupils to get over hurdles in manipulating equipment within a limited time, but it hinders pupils' minds on inquiring and meaning making. Research on practical work often uses the "cookbook" metaphor to describe this prescriptive instruction (Fig. 3.9).

Considering the ICT integration, extra instructions needed to handle the necessary software might further reinforce the prescriptive nature of ICT-enhanced experimentation/modelling activities. These instructions can unintendedly come to dominate the activity, although there are practical ways to get around this. For example, after a cookbook phase for learning to manipulate the tools should come an inquiry phase in which pupils themselves have to make decisions on how to use the new tools in their investigations.

Limited Preparation Time and Limited Curriculum Time In many countries including Slovakia and Vietnam, teachers are pressed to teach all of the content standards within a tightly structured, explicitly expressed syllabus (Woolnough 2001). In the Netherlands, the school science curricula are rather overloaded with standardised

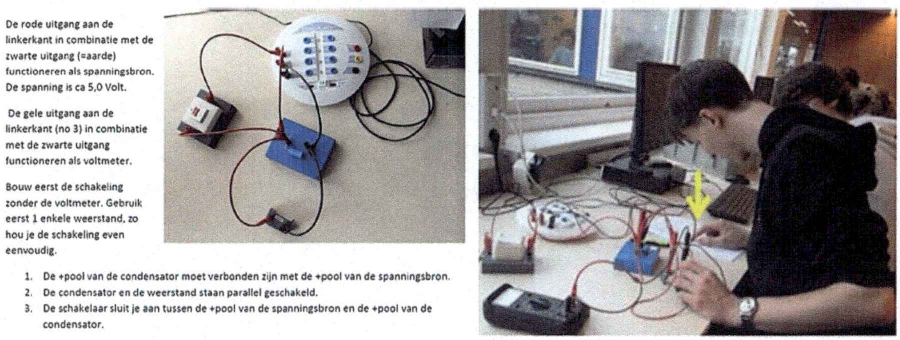

(a) (b)

Fig. 3.9 In the photograph (**b**), a pupil was following a cookbook instruction (**a**) to set up the experiment using sensors and the computer

requirements, too; except that the way to attain these requirements is mostly left open to the teacher. In such constrained circumstances, teachers tend to get through the content (Bencze and Hodson 1999) rather than engaging pupils in investigations with the ICT tools. This is because these investigations require ample time and aim at not only conceptual learning but also other goals. Incorporating the ICT tools in school science might add extra problems to time-constraint situations, considering that teachers need extra time, effort, and/or training to learn to handle the ICT tools and to practise their use. Logistics of organising the use of hardware and software cause extra preparation time for teachers. The constraints on preparation time and curriculum time can be a factor explaining for the fact that: although the ICT tools have many innovative features that stimulate pupils' authentic inquiry practices, really authentic inquiry learning is limited to a few special projects, for example, in Dutch schools, once in junior secondary and once in senior secondary.

Conclusions About Challenges of Technology in Inquiry-Based Teaching of Physics

The science education community mostly agrees about the relevance of using ICT, IBSE, and their integration: ICT in IBSE for pupils exercising inquiry practices, acquiring inquiry skills, and understanding scientific inquiry. However, ICT in IBSE is still very much underused and applied at a relatively small scale in most countries. When it is used, the use often lacks the basic characteristics of inquiry.

Although integration of ICT into IBSE is relevant, it is cognitively and practically challenging for both students and teachers. ICT-enhanced inquiry-based strategies have proved their potential, but not their general efficacy in the hands of average teachers. Science teachers need effective training and practical guidelines to handle the complexity of the ICT-enhanced, inquiry-based activity without reducing it to simple "cookbook recipes". Furthermore, integration of ICT into IBSE needs sufficient time to be faithfully implemented. An actual impact of ICT in IBSE teaching on pupils needs both a longer period of time and consistent incorporation of ICT in IBSE in regular teaching.

3.5 Development of a Short and Effective Course for Teachers on Technology in Inquiry-Based Teaching of Physics

Aim and Research Questions

Factors involved in the underuse of ICT and IBSE in teaching practice are among others (a) limited curriculum time and limited teacher preparation time; (b) mismatches of the IBSE goals with commonly used lesson materials, teaching methods, and assessment and examination (e.g. prescriptive nature of materials and methods, predominance of content over inquiry goals); and (c) insufficient teacher preparation and training on integrating ICT into IBSE. All of these factors need to be changed consistently and in concert to realise proper incorporation of ICT and IBSE into a classroom where *manipulation of equipment and software* is turned to *manipulation of ideas and concepts* for knowledge generation and validation.

Within our recent research project (Tran 2016), we focussed on preparation and training of science teachers on ICT in IBSE teaching and developed an effective and relatively short course for student teachers and teachers with diverse teaching experience. The present research confined the ICT in IBSE teaching to (a) three Coach tools, data logging with sensors, video measurement, and dynamical modelling, and (b) the use of these tools to support inquiry by pupils. In an ICT in IBSE activity, the pupils should have some role not only in executing the experiment/model but also to some extent in formulating research questions, designing the experiment/model, and interpreting the results. We developed a short course so that it can be accommodated within typical overloaded teacher-education programmes or adopted as an in-service course. Furthermore, educational theories and products, such as our ICT in IBSE course, do not always travel well as educational and cultural contexts in different schools and countries can be very different. That is why the present research included three case studies in the Netherlands, Slovakia, and Vietnam and in pre- and in-service teacher education. That way we could test the transferability of our course design and the generalisability of the pedagogical principles at the basis of this design.

The aim of the present research was twofold. First, the objective was to design a short course, which—with some adaptations—will be effective in widely different educational settings. Second, this research was to investigate the validity of pedagogical principles, which were used to guide (a) the design, implementation, evaluation, and optimisation of the course and (b) the extent to which the course can be adjusted to the different settings in the Netherlands, Slovakia, and Vietnam. The pedagogical principles are at a higher level of abstraction and intended to be generalisable across educational and cultural contexts.

The design research approach was applied as it can provide guidelines and scientific reasoning for such a research and design process, which was guided by the two research questions.

First, what are characteristics of an effective, short course for Dutch student teachers to learn to apply the ICT tools in IBSE?

Second, to what extent is the course applicable in different educational and cultural contexts of pre- and in-service teacher education in different countries (i.e. the Netherlands, Slovakia, and Vietnam)?

Course Design and Research Design

Objectives of the ICT in IBSE Course The general aim of the course was elaborated into the four objectives as follows:

1. Awareness objective: participants become aware of educational benefits of the ICT tools in science education.
2. ICT mastery objective: participants master skills to operate the ICT tool.
3. ICT in IBSE objective: participants can design, implement, and evaluate an ICT in IBSE lesson.
4. Motivation objective: participants are motivated to continue studying the ICT tools and trying out ICT in IBSE lessons with pupils.

The ICT in IBSE objective (3) was considered as the main objective of the course. In order to reach this objective, participants had to achieve a certain minimum level of mastery of the ICT tools (2). The awareness objective (1) and motivation objective (4) were aimed at the course's long-term effects on participants' teaching practice. The Coach platform for data logging, video measurement, and modelling was used together with available support materials (i.e. Coach introductory, tutorial, and exemplary activities) and materials that we developed (i.e. forms for designing and self-evaluating the ICT in IBSE lesson).

Pedagogical Principles Underlying the ICT in IBSE Course The literature on design research and on professional development of teachers led us to the following pedagogical principles as the basis for (re)designing, evaluating, and optimising the ICT in IBSE course:

1. *One theory-practice cycle*: participants are required to go through at least one complete cycle of designing, implementing, and evaluating an ICT in IBSE lesson within the course. Participants will apply the IBSE theory in a design for an ICT in IBSE lesson, which they will also try out in the classroom, self-evaluate, and report in the final session of the course.
2. *Distributed learning*: participants study in live sessions and carry out individual assignments in between the sessions with the support materials and in consultation with the course instructor. Learning time is distributed between live sessions and individual assignments but is also carefully distributed over a longer period to provide opportunities for a well-planned try-out in a real classroom.
3. *Depth first*: participants are introduced to the possibilities of the three tools after which they specialise in only one ICT tool. Learning time is prioritised for an in-depth study and application of one tool (one-tool specialisation) rather than broad study of all three tools at a more superficial level, so depth first—breadth later.
4. *Ownership of learning*: participants have freedom to select what to learn and how to learn it, using the course scenario and support materials in order to achieve the course objectives. The individual participants pursue their self-tailored learning process in which they make their own choices regarding the tool, the grade level, topic, and activity for their ICT in IBSE try-out with pupils.
5. The four objectives, four pedagogical principles, and support materials together form the general design of the ICT in IBSE course.

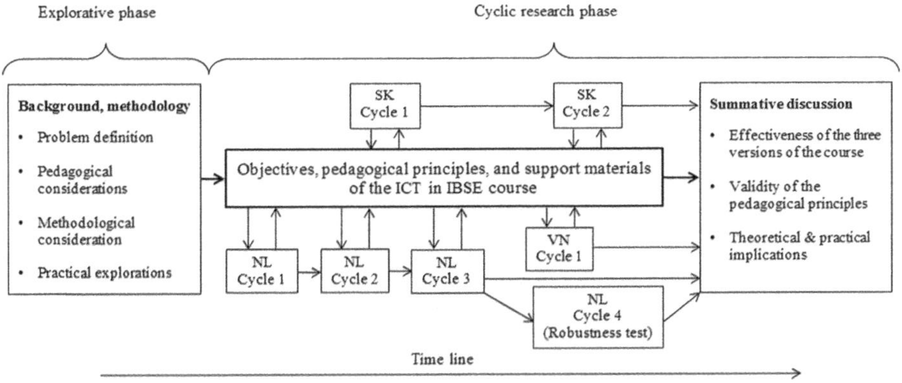

Fig. 3.10 Design and research process including an explorative phase and a cyclic-research phase with three case studies in the Netherlands (NL), Slovakia (SK), and Vietnam (VN)

Evaluation of the ICT in IBSE Course Through Three Case Studies To examine the effectiveness of the ICT in IBSE course, validity and generalisability of the pedagogical principle, and transferability of the course design in different education contexts, we conducted the Dutch, Slovak, and Vietnamese case studies (Fig. 3.10). These three case studies were related; the Dutch case study was the earliest and most extensive, followed by the Slovak case study, and then the Vietnamese case study. All three case studies (a) concerned the same questions about implementation of the pedagogical principles and course design, usefulness of the support materials, and attainment of the course objectives, (b) applied the same evaluation framework, and (c) used the same instruments for data collection and analysis. The course evaluation was guided by two main questions:

(A) To what extent were the four pedagogical principles implemented as intended?
(B) To what extent did the ICT in IBSE course achieve its four objectives?

Question B involves the evaluation of the effects of the course on participants, which resulted from actual implementation of the pedagogical principles, the course design, and the support materials. The evaluation of this actual implementation was guided by Question A and based on a comparison between (a) the intended course programme and (b) the actual activities of participants during the course. To evaluate attainment of the course objectives, we first operationalised performance levels for each objective. The definition of these levels was based on theoretical considerations and aligned with time-constraint conditions of the course. After that, we collected data, compared the data-analysis outcomes with the predefined levels of the course objectives, and concluded which level(s) of each objective the participants achieved.

This evaluation framework includes instruments for data collection and analysis (i.e. pre-course, post-course, and follow-up questionnaires; observations and video recording of live sessions and classroom try-outs; participants' ICT in IBSE lesson plans and self-evaluation reports of the classroom try-outs; computer performance test for each tool; the inquiry-analysis inventory; and the communication records).

With these instruments, data were collected from a variety of sources and by different data collectors (i.e. the researcher, the course instructor, course participants). Accordingly, we could record both intended and possibly unintended outcomes as the course was implemented. Most outcomes were evaluated by more than one instrument thus allowing for data triangulation.

In the Dutch case study, we further operationalized the pedagogical principles in the initial scenario of the ICT in IBSE course. With "scenario", we mean the programme of the course and all instructor and participant activities and assignments. After that, we implemented and evaluated the course with 40 physics/chemistry student teachers spread over four sequential cycles. Among these four cycles, Cycles 1 and 2 were for fine-tuning of the course scenario. The course evaluation (Questions A and B) and experiences with the course in Cycle 1 (including what did work, what did not work, and why) suggested revisions of the initial scenario. These revisions were aimed at more faithful implementation of the course in Cycle 2 and with respect to many factors such as diversity of participants' background and ability, school schedules, and curriculum time for ICT in IBSE try-outs. Likewise, the Cycle 2 evaluation was guided by the objectives and pedagogical principles and resulted in further optimisation of the course scenario. We achieved faithful implementation of the four principles in Cycle 3. Consequently, in this cycle, the summative effects of the Dutch version of the ICT in IBSE course were evaluated, and only minor suggestions were made for further optimisation. The robustness of the course design and the ecological validity of the pedagogical principles were tested in Cycle 4 under routine implementation conditions without the extra support of the researcher.

The new understanding of how the course was developed and why it was effective (Dutch context) together with the basic course design (including course objectives, pedagogical principles, and support materials) enabled the tailoring of local versions of the ICT in IBSE course in different contexts. The ICT in IBSE course was adapted and tested in (a) two cycles with 66 physics/biology/chemistry teachers with diverse teaching experience (1 to 33 years of teaching) in Slovakia. The two cycles of the Slovak course were already in routine implementation conditions without the direct participation of the researcher. The ICT in IBSE course was adapted and tested in one cycle in Vietnam with 22 master students in physics education, who either had taught for 2–9 years or came straight from a Bachelor teacher-education programme. Evaluations of the three local versions of the course enabled us to draw conclusions about (a) the extent to which the four objectives can be attained, (b) the validity and generalisability of pedagogical principles, (c) the transferability of the course design, and (d) the practical relevance of the course. These evaluations led us to new understanding of the extent to which the pedagogical principles can guide the fine-tuning of the basic design of the ICT in IBSE course to varying boundary conditions.

3.6 Findings, Discussion, and Conclusions

The Dutch case study resulted in an improved and successful course scenario in which the Dutch participants achieved the course objectives also when the course was taught under routine conditions (Cycle 4). A typical reaction of one of the participants was:

> Very relevant, I am happy that the course was offered and that I took it. In the teacher-education programme, we learn a lot of theory but for the implementation of it, I feel rather on my own, with very little guidance. The ICT in IBSE course was very hands-on (though the theory was also presented), gave me the opportunity to learn with and from others (knowledgeable, friendly experts and other (beginning) teachers). I got the guidance I needed to develop an activity and implement it, and the feedback after completion. The duration was fine. I strongly advise the training to other student teachers

Figure 3.11 visualises the optimised scenario, illustrates the specific operationalisation of the pedagogical principles in the Dutch context, and describes the time distribution for the three live sessions and two assignments and the plan on when to use which support materials and data-collection instruments. For more detail about the programme for the live sessions, the requirements for the individual assignments, and the support materials, see Tran (2016) and visit http://cma-science.nl/ict-ibse-course.

The iterative evaluation and refinements of the Dutch course confirmed the validity of the pedagogical principles in the Dutch context. The support materials proved necessary and useful for the sufficient implementation of the pedagogical principles and the satisfactory attainment of the course objectives. To conclude, the four course objectives, the four pedagogical principles, and the optimised scenario with the support materials establish the core characteristics and basic design of an effective short ICT in IBSE course for Dutch student teachers. Findings from the iterative refinement of the course show that fine-tuning the distribution of time and individual assignments is crucial as far as distributed learning is concerned. Direct, personalised support (in live

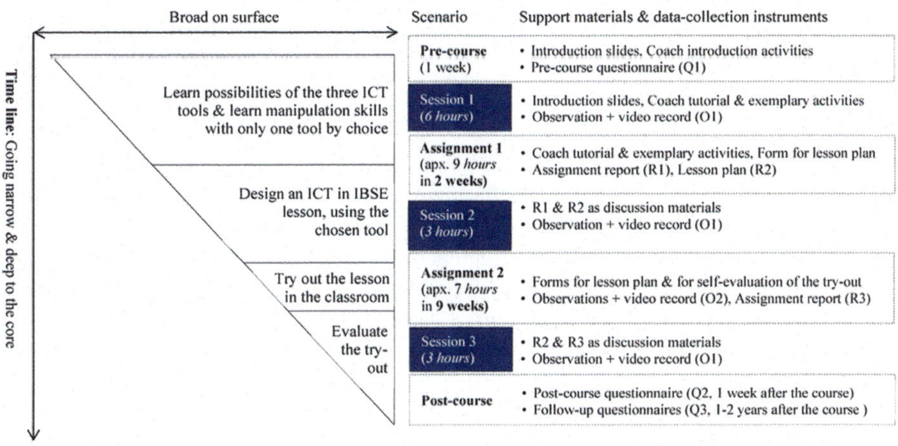

Fig. 3.11 The optimised scenario of the ICT in IBSE course in the Dutch context

Table 3.1 Differences among Dutch, Slovak, and Vietnamese contexts for the course

	Dutch context	Slovak context	Vietnamese context
Participants			
• Number	Forty student teachers	Sixty six experienced teachers	Twenty two master students
• Background and age	Mix of fresh master and second-career graduates; physics and chemistry Age: 23–55	Mix of physics, chemistry, biology, and geography teachers Age: 23–55	Mix of fresh graduates and experienced teachers, physics Age: 23–31
• Teaching experience	1–5 years, 83% at first-year teaching	1–33 years, 19 years on average	0–9 years, 23% with no teaching
• Entrance level of ICT skills	Moderate	Low	Low
Scheduling requirements			
• Programme	Post-graduate teacher education	Accredited professional development	Master in physics education
• Total study time of the course	28 h	40 h	60 h
• "Spread" of the course	11 weeks	15 weeks	5 weeks
Teaching conditions			
• Availability of the ICT tools	Sufficient *Available in most schools*	Limited	None
• Pupil experience with ICT	Sufficient *Pupils have ever used Coach/similar software*	Insufficient *ICT is starting to be introduced*	None
• Pupil experience with IBSE	Insufficient *Certain experience with laboratory but less with IBSE*	Poor *A little experience with laboratory but very limited with IBSE*	None
• IBSE in curriculum	Explicit and required	Starting to be required	In new 2016 curriculum
• Teacher autonomy	High	Moderate	Low

meetings and/or via emails) and sense of direction (via explicit support framework plus assignment tracking and stimulation) are crucial factors to ensure effectiveness of independent learning, especially the quality of the ICT in IBSE lesson plan. These factors create a balance between much freedom of choice and appropriate guidance, which is essential to establish ownership of learning.

The Dutch, Slovak, and Vietnamese contexts for the ICT in IBSE course were different in many aspects (e.g. scheduling requirements, school conditions, and characteristics of participants) (Table 3.1). First, the Dutch course was limited to 12 contact hours out of 28 h of total study time, but it was spread over 11 weeks. The Vietnamese course was compressed in 5 weeks, but 30 h out of total 60 study hours were scheduled for live activities. The Slovak case had the least constraints, regarding both contact hours (25 out of total 40 study hours) and "spread" of the course (15 weeks). Second, the Dutch school conditions (e.g. curriculum time, teacher preparation time, national examinations, pupils' experience with ICT and IBSE, availability of equipment and software) were not excellent but sufficient. Meanwhile, the Slovak school conditions were insufficient, and the Vietnamese conditions were very poor. Third, the Vietnamese and Slovak participants were experienced teachers, but their ICT mastery entrance level was low. The Dutch participants had more experience with the ICT tools and felt more free to decide their own lesson objectives and teaching methods. However, they lacked teaching experience, especially classroom management skills. Vietnamese teachers work in an education system with a strong hierarchical culture and much less autonomy than in the Dutch system. Lessons are teacher-centred, and there is no tradition of open learner investigations in secondary school and teacher education. All three groups of participants lacked practical experience with inquiry teaching with or without ICT, so ICT in IBSE teaching was challenging for them. For all three versions of the course, diversity of participants and time constraints were challenging contextual factors.

Across the three case studies, the *awareness* and *motivation objectives* of the ICT in IBSE course were achieved as expected. The participants could enumerate relevant benefits of the ICT tools. They devised plans and actually continued studying the ICT tools and teaching ICT in IBSE lessons after the course. About the *ICT-mastery objective*, all three groups of participants were able to operate the Coach tool fluently after the course. Compared with the Dutch participants, the Vietnamese participants attained a higher mastery level for the chosen tool, and the Slovak participants achieved a similar ICT mastery but with all three ICT tools. This shows effectiveness of the many more contact hours with direct, personalised support scheduled for the ICT-mastery objective to compensate for the low ICT entrance of the Slovak and Vietnamese participants.

About the *ICT in IBSE objective*, all three groups of participants were able to design and realise acceptable ICT in IBSE lessons considering their teaching conditions and their inexperience with inquiry teaching with ICT. The Dutch participants could design and realise better ICT in IBSE lessons than the Slovak and Vietnamese participants. Many Dutch participants were able to engage pupils in designing experiments or models and predicting and interpreting results as expected. Meanwhile, the Slovak and Vietnamese participants focused too much on pupils' execution of experiments or models (manipulation of equipment and software) and did not sufficiently involve pupils in moving back and forth between the physical and theoretical worlds (manipulation of ideas and concepts). Most Slovak and Vietnamese participants intended to take control over the entire classroom activity through plenary systematic explanations and/or prescriptive worksheets for the group work. In contrast, in half of the Dutch ICT in IBSE lesson plans, pupils were required to take a larger role in conception, planning, and interpretation of the experiment/model in more open inquiry patterns. This shows a clear difference in teacher/pupil centeredness and education culture among the three countries.

Although familiar with theory of IBSE, all three groups of participants had trouble to operationalise real inquiry in lesson plans and even more so in the classroom. There were many deviations between intended and actual ICT in IBSE lessons, and these resulted from reasons such as shortcut of intended inquiry opportunities, tasks that were too demanding, overambitious timing, and ineffective communication with pupils. However, Dutch, Slovak, and Vietnamese participants were able to identify the shortcomings in their ICT in IBSE lessons and suggest relevant revisions of their lesson plans for future use. To conclude, the basic design of the ICT in IBSE course was *effective, practical, and transferable* in the different educational and cultural contexts of pre- and in-service teacher education in different countries. The course can cater to diverse groups of teachers and teacher-education programmes, and it fits into time-constraint conditions. For all three cases, the ICT in IBSE course achieved its objectives to the predetermined acceptable level, except that for the ICT in IBSE part, there was still much room for improvement.

Considering the issue of teachers learning to teach by inquiry, we prepared and expected our course participants to get their first experience with inquiry teaching with ICT. The theory-practice cycle was valuable to make them *more aware* of what IBSE involves, of what are differences between guided versus open inquiry, and of how to involve pupils in planning and interpretation of an experiment. It was concluded that the educational and cultural system influences teachers' perception and implementation of inquiry-based teaching with ICT. This results in different typical patterns of ICT in IBSE in different countries. The analysis of the lesson plans and classroom try-outs using the inquiry-analysis inventory revealed considerable *inconsistency between inquiry objectives and activity specifications* and noticeable *deviations between intended and actual IBSE lessons*. These are persistent problems, which have been reported worldwide (Abrahams and Millar 2008; Abrahams and Reiss 2012; Tamir and Lunetta 1981). Many teachers do have problems to operationalise inquiry in the classroom, even in countries like the United Kingdom and the United States where inquiry has been emphasised in the curriculum for a long time. Research findings from the Vietnamese case study shed light on challenges of and potential solutions to the application of IBSE in a hierarchical education culture. Obviously, the ICT in IBSE course under the time constraints does not push its participants far enough yet in the direction of inquiry teaching with ICT. Participants' achievement through the course is a starting point; more theory-practice cycles are needed to bring them further in such ICT in IBSE direction.

In the present research, the applied pedagogical principles proved to be valid in providing not only the framework for implementing, evaluating, and optimising the course in a specific context but also guidelines for effective adaptation of the course to varying boundary conditions. When adapting the course to a different context, the "one theory-practice cycle" principle should not be changed. Instead, the "depth-first" and "distributed learning" principles can be adjusted by the course instructor to some extent to the specific context, considering the entrance level and other characteristics of the participants and the scheduling requirements. The "ownership of learning" principle has to be enabled to provide a dial for participants to self-tune the course to their own interest and ability. The adjustment with distributed learning and depth first makes the first flexible phase: ICT mastery, which can be lengthened (Slovak case) and compressed (Vietnamese case) in order to compensate for the low

ICT entrance, accommodate diversity of the participants, and align their activities, assignments, and efforts with the intended attainment of the ICT mastery objective. Such ICT mastery attainment is necessary for the participants to be able (a) to design and teach the ICT in IBSE lesson and (b) to continue studying and using the ICT tools after the course. Among the four course objectives, the ICT-mastery objective can be achieved in a compressed course with sufficient contact hours, whereas the learning with respect to the ICT in IBSE objective needs to be distributed sufficiently to allow for a well-planned and mature lesson plan and curriculum time for classroom try-out. The support materials proved necessary, useful, and robust in different contexts. This finding suggests that it is not always necessary to develop materials locally to have effective educational innovations. Instead, with certain adaptations, one can use existing materials.

Reflections on the Findings and Methods

Based on our positive experiences with one theory-practice cycle, we think that this principle should be wider applied in teacher education. Would it be possible to identify a small number of core practices and have student teachers go through one theory-practice cycle for each? For example, the study of formative assessment could be followed by classroom practice with embedded formative assessment and feedback. Regarding the depth-first principle, deeper understanding of one ICT tool has surplus value compared to partial understanding of all three tools, and it leads to better transfer to the whole ICT environment (breadth later). This further suggests the application of the depth-first principle as part of a solution for content overload in teacher-education programmes.

Regarding the learning of ICT skills, collective practice of ICT skills in small groups is more effective than either individual practice at home or the practice under plenary step-by-step instructions to the whole class. Personalised, direct support from the course instructor and peers is essential for participants to get over initial hurdles of learning a new tool and to troubleshoot "technological content knowledge" (TCK) problems. Beyond the basic manipulation skills, TCK problems involve:

- Advanced features of the ICT tool (e.g. how to add the empirical graph to the modelling activity and compare it with the modelling graph; how to make control bars to change constants and initial values of variables of the model)
- Participants' understanding of the phenomenon (i.e. concepts, laws, and events) to be experimented or modelled with the ICT tool and general experimentation/modelling skills
- Participants' knowledge of mental model behind the digital tool that enables (a) collection and modelling of information about the phenomenon (sampling, digitalisation, numerical iteration) and (b) representation, analysis, and processing of the collected/modelled data.

To troubleshoot TCK problems independently, participants need to understand how the ICT system works and to be confident and committed in searching and trying out solutions. Taking the computer performance test can be a first step to learn such troubleshooting skills.

At the beginning of the present research, we defined the unique objectives of the course to be developed and a clear view about design criteria. Based on these objectives and criteria, we chose the design research approach to develop, evaluate, and optimise the course as an educational product through research, and this approach worked well. We started this research and design project with the pedagogical principles and concluded it with these principles as the core of the basic design of the ICT in IBSE course. These principles can be considered as independent, validated educational products, which teacher educators can "buy into" and use for broader aims than only ICT in IBSE integration. Pedagogical principles establish the theoretical model underlying the course design; provide guidelines and structure to the (re)design, implementation, evaluation, and optimisation process; and help to communicate the design research to others. The role of pedagogical principles in design research is indeed essential. Moreover, in our design research, we incorporated (a) a "robustness test" step to try out the course under routine conditions and (b) a "generalisability and transferability testing" step to try out the course in different programmes or even countries. We achieved successful outcomes with these steps. Consequently, we strongly recommend robustness and generalisability/transferability tests as part of design research.

Main limitations of the present research were that it only measured the quality of one theory-practice cycle, that it did not have an opportunity to measure the further development of participants in later ICT in IBSE activities in their classrooms, and that it did not measure pupil results. Learning to teach the IBSE way takes a longer trajectory than this course, and so our measurements only show the start of the participants' development. For the same reason, there was no point in measuring pupil achievement, as improvement of their inquiry skills would only become visible after a series of lessons rather than one lesson.

It is common practice that each teacher-education programme invents its own wheels. Our research outcomes through the development/adaptation, implementation, and evaluation of the ICT in IBSE course in the Netherlands, Slovakia, and Vietnam indicate that with careful design and well-chosen pedagogical principles, courses and other educational products could be fine-tuned and shared. The fourth and fifth Dutch course and the second Vietnamese course were implemented by the local course instructors without any involvement of the researcher. The third Slovak course is planned within a new large-scale national project, which is aimed at implementation of ICT tools across science subjects in Slovakia. The Dutch ICT in IBSE course (implemented with five batches already) is unique, since it is the only teacher-education course offered by several Dutch universities together, as far as we know. The Vietnamese ICT in IBSE course is unique as it is the only master course of which design and materials were entirely developed outside the university. These institutionalisations do not happen often for general educational projects. These show not only the *practical relevance of the basic design* of the ICT in IBSE course for different educational and cultural contexts of pre- and in-service teacher education but also suggest the possibility to have *more-productive standardisation* among teacher-education courses. Education should be less political-driven and more expert-driven!

References

Abrahams, I., & Millar, R. (2008). Does practical work really work? A study of the effectiveness of practical work as a teaching and learning method in school science. *International Journal of Science Education, 30*(14), 1945–1969.

Abrahams, I., & Reiss, M. J. (2012). Practical work: Its effectiveness in primary and secondary schools in England. *Journal of Research in Science Teaching, 49*(8), 1035–1055.

Barton, R. (2004). *Teaching secondary science with ICT*. Maidenhead, England: McGraw-Hill Education.

Beichner, R. J. (1996). The impact of video motion analysis on kinematics graph interpretation skills. *American Journal of Physics, 64*(10), 1272–1277.

Bencze, L., & Hodson, D. (1999). Changing practice by changing practice: Toward more authentic science and science curriculum development. *Journal of Research in Science Teaching, 36*(5), 521–539.

Brasell, H. (1987). The effect of real-time laboratory graphing on learning graphic representations of distance and velocity. *Journal of Research in Science Teaching, 24*(4), 385–395.

Braund, M., & Reiss, M. (2006). Towards a more authentic science curriculum: The contribution of out-of-school learning. *International Journal of Science Education, 28*(12), 1373–1388.

Davis, E. A., Petish, D., & Smithey, J. (2006). Challenges new science teachers face. *Review of Educational Research, 76*(4), 607–651.

De Jong, T., & van Joolingen, W. R. (1998). Scientific discovery learning with computer simulations of conceptual domains. *Review of Educational Research, 68*(2), 179–201.

Dewey, J. (1910). Science as subject-matter and as method. *Science, 31*, 121–127.

Duschl, R. A., Schweingruber, H. A., & Shouse, A. W. (2007). *Taking science to school: Learning and teaching science in grades K-8*. Washington, DC The National Academies Press.

Edelson, D. C. (1998). Realising authentic science learning through the adaptation of scientific practice. In B. J. Fraser & K. G. Tobin (Eds.), *International handbook of science education* (Vol. 1, pp. 317–331). Great Britain: Kluwer.

Ellermeijer, A. L., Landheer, B., & Molenaar, P. P. M. (1996). Teaching mechanics through interactive video and a microcomputer-based laboratory. In R. F. Tinker (Ed.), *Microcomputer-based Labs: Educational Research and Standards* (pp. 281–290). Amsterdam: Springer-Verlag.

Gaskell, P. J. (1992). Authentic science and school science. *International Journal of Science Education, 14*(3), 265–272.

Gopnik, A., Meltzoff, A. N., & Kuhl, P. K. (1999). *The scientist in the crib: What early learning tells us about the mind*. New York: Perennial.

Gröber, S., Klein, P., & Kuhn, J. (2014). Video-based problems in introductory mechanics physics courses. *European Journal of Physics, 35*(5), 055019.

Heck, A. (2007). Modelling intake and clearance of alcohol in humans. Paper presented at the International Conference on Technology in Mathematics Teaching (ICTMT8), Czech Republic.

Heck, A. (2009). Bringing reality into the classroom. *Teaching Mathematics and its Applications, 28(4)*, 164–179.

Heck, A., & Ellermeijer, A. L. (2009). Giving students the run of sprinting models. *American Journal of Physics, 77*(11), 1028–1038.

Heck A, Uylings P. (2005). *Yoyo Joy*. In: F. Olivero & R. Sutherland (Eds.) Technology in Mathematics Teaching, Proceedings of ICTMT7 Vol. 2; Bristol: University of Bristol; pp. 380–387.

Heck, A., Ellermeijer, A. L., & Kedzierska, E. (2009a). Striking results with bouncing balls. Paper presented at the Physics Curriculum Design, Development and Validation, Nicosia, Cyprus.

Heck, A., Kedzierska, E., & Ellermeijer, A. L. (2009b). Design and implementation of an integrated computer working environment for doing mathematics and science. *Journal of Computers in Mathematics and Science Teaching, 28*(2), 147–161.

Heck, A., Uylings, P., Kędzierska, E., & Ellermeijer, A. L. (2010). *Cross-disciplinary, authentic student research projects*. Paper presented at the SMEC 2010 conference: Inquiry-based learning: Facilitating authentic learning experiences in science and mathematics, Dublin, 40–45.

Hodson, D. (2009). *Teaching and Learning about Science: Language, Theories, Methods, History, Traditions and Values*. Rotterdam: Sense Publishers.

Hofstein, A., & Lunetta, V. N. (2004). The laboratory in science education: Foundations for the twenty-first century. *Science Education, 88*(1), 28–54.

Jeskova, Z., Kires, M., McLoughlin, E., Finlayson, O., Ottander, C., & Ekborg, M. (2015). In-service and pre-service teacher education in IBSE – The ESTABLISH approach. In C. Fazio & R. M. Sperandeo Mineo (Eds.), Proceedings of the GIREP-MPTL International Conference on Physics Education (pp. 811–818). Palermo: Università degli Studi di Palermo.

Kearney, M., & Treagust, D. F. (2001). Constructivism as a referent in the design and development of a computer program using interactive digital video to enhance learning in physics. *Australasian Journal of Educational Technology, 17*(1), 64–79.

Kirschner, P. A., Sweller, J., & Clark, R. E. (2006). Why minimal guidance during instruction does not work: An analysis of the failure of constructivist, discovery, problem-based, experiential, and inquiry-based teaching. *Educational Psychologist, 41*(2), 75–86.

Laws, P., & Pfister, H. (1998). Using digital video analysis in introductory mechanics projects. *The Physics Teacher, 36*(5), 282–287.

Minner, D. D., Levy, A. J., & Century, J. (2009). Inquiry-based science instruction—what is it and does it matter? Results from a research synthesis years 1984 to 2002. *Journal of Research in Science Teaching, 47*(4), 474–496.

Newton, L., & Rogers, L. (2001). *Teaching science with ICT*. London: Continuum.

NRC (National Research Council). (2012). *A Framework for K-12 Science Education: Practices, Crosscutting Concepts, and Core Ideas*. Committee on a Conceptual Framework for New K-12 Science Education Standards. Board on Science Education, Division of Behavioural and Social Sciences and Education. Washington, DC: The National Academies Press.

OECD (Organisation for Economic Co-operation and Development). (2006). *Evolution of student interest in science and technology studies: Policy report*. Retrieved from http://www.oecd.org/science/sci-tech/36645825.pdf

Ogborn, J. (2014). Curriculum development in physics: Not quite so fast. In M. F. Tasar (Ed.), *Proceedings of The World Conference on Physics Education* (pp. 39–48). Istanbul: Pegem Akademi.

Paavola, S., Lipponen, L., & Hakkarainen, K. (2004). Models of innovative knowledge communities and three metaphors of learning. *Review of Educational Research, 74*(4), 557–576.

Rogers, L., & Twidle, J. (2013). A pedagogical framework for developing innovative science teachers with ICT. *Research in Science & Technological Education, 31*(3), 227–251.

Sokoloff, D. R., Laws, P. W., & Thornton, R. K. (2007). Real Time Physics: Active learning labs transforming the introductory laboratory. European Journal of Physics, 28(3), S83–S94.

Tamir, P., & Lunetta, V. N. (1981). Inquiry-related tasks in high school science laboratory handbooks. *Science Education, 65*(5), 477–484.

Tran, T. B. (2016). *Development of a course on integrating ICT into inquiry-based science education*. Doctoral thesis, VU University Amsterdam.

van Buuren, O. (2014). *Development of a modelling learning path.* Doctoral thesis, University of Amsterdam, Amsterdam.

van Buuren, O., Uylings, P., & Ellermeijer, A. L. (2010). Towards a learning path on computer modelling. In D. Raine, C. Hurkett, & L. Rogers (Eds.), *Proceedings of the GIREP-EPEC & PHEC 2009 International Conference* (Vol. 1, pp. 110–125). Leicester: Lulu/The Centre for Interdisciplinary Science.

van den Berg, E. (2013). The PCK of laboratory teaching: Turning manipulation of equipment into manipulation of ideas. *Scientia in Educatione, 4*(2), 74–92.

Velanova, M., Demkanin, P., Gergelova, B., & Demkaninova, D. (2014). The experienced physics teacher and her first experience with data-logger. *In Proceedings of INNODOCT/14 conference: Strategies for Education in a New Context* (pp. 142–157). Spain: Editorial Universitat Politècnica de València.

Woolnough, B. E. (2001). Of 'knowing science' and of 'doing science': A reaffirmation of the tacit and the affective in science and science education. *Canadian Journal of Math, Science & Technology Education, 1*(3), 255–270.

Zollman, D., & Fuller, R. (1994). Teaching and learning physics with interactive video. *Physics Today, 47*, 41–47.

Chapter 4
Sharing LHC Research and Discovery with High School Students

Marcia Begalli and Uta Bilow

Abstract Research institutes and universities around the world invite students and their teachers for a day-long programme to experience life at the forefront of basic research. These *International Masterclasses* (www.physicsmasterclasses.org) give students the opportunity to be particle physicists for a day by analysing real data from CERN's Large Hadron Collider (LHC). The project attracts each year more than 13,000 high school students from 46 countries.

In the *International Masterclasses*, high school students work with real data collected by the experiments at the LHC. The programme bridges the gap between science education at school and modern scientific research. Participants can explore the fundamental forces and building blocks of nature and are informed about the new age of exciting discoveries in particle physics, e.g. the discovery of the Higgs boson. Moreover, they can actively take part in cutting-edge research and improve their understanding in science and the scientific research process. The programme offers authentic experience and adds valuable experiences to physics education at school, thus stimulating the students' interest in science.

4.1 Introduction

The Large Hadron Collider (LHC) is the world's largest and most powerful particle accelerator. The 27 km ring is situated next to Geneva, Switzerland, at the CERN site. Scientists at the LHC are conducting a broad research programme, which aims at fundamental questions on origin and structure of our world. By tracing processes between elementary particles, they shed light on the development of the universe 10^{-12} s after the Big Bang (see Fig. 4.1).

M. Begalli
State University of Rio de Janeiro, Rio de Janeiro, Brazil

U. Bilow (✉)
Technische Universitaet Dresden, Dresden, Germany
e-mail: uta.bilow@tu-dresden.de

M. Pietrocola (ed.), *Upgrading Physics Education to Meet the Needs of Society*,
https://doi.org/10.1007/978-3-319-96163-7_4

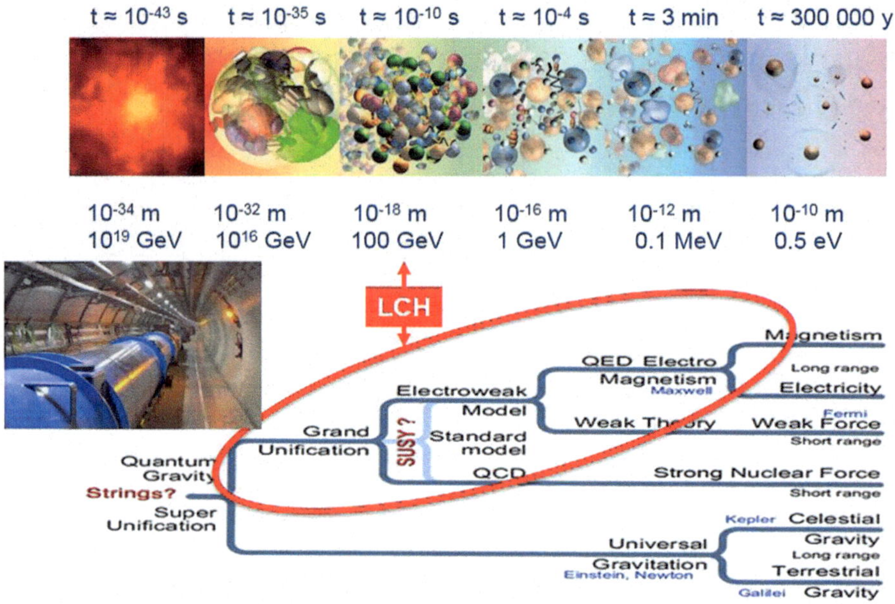

Fig. 4.1 Evolution of the universe from the Big Bang to the present: the LHC looks at $<10^{-10}$ s after the Big Bang

Results of the experiments at the LHC are followed with great interest, not only by physicists but also by the general public. The discovery of the Higgs boson at the LHC in summer 2012 led to a huge media echo and large public interest. Although particle physics is one of the most fascinating and emerging fields in science, it is often not covered in curricula and school lessons. There are several reasons for this situation: Introducing particle physics requires many new concepts and perceptions, e.g. strong charge, weak charge, messenger particles, and thus a large number of new terms. In addition, there is no strong connection between the world of elementary particles and everyday experience. In contrast to that, particle physics has inherent strength and attractive features and thus offers several chances to successfully introduce high school students to the subject:

- Unique experimental setups with highly complex technique create superlatives, e.g. one of the coldest places on earth, largest superconducting installation, biggest operational vacuum system
- Fundamental questions on origin and development of our universe, elementary building blocks which make up the world around us and the fundamental forces between them
- Fascinating terms, e.g. Big Bang, antimatter
- Strong presence in the media

Fig. 4.2 High school students at a masterclass

4.2 Particle Physics Masterclasses

International Masterclasses[1] is a programme that has been created to offer high school students (aged 15–19) the chance to explore this field of cutting-edge physics by working with recent, authentic data from experiments at the LHC (Bilow and Kobel 2014; Bardeen et al. 2014). In this way, *International Masterclasses* bridges the gap between science education at school and modern scientific research. The basic idea of the annual programme is to let students work as much as possible like real scientists in a format that is called masterclass (Fig. 4.2).

As in a masterclass in the arts, high school students work with a particle physicist as an expert on the subject particle physics data analysis. The students get invited to a nearby research institute or a university to be "scientists for 1 day" and to experience life at the forefront of basic research by analysing real data from LHC experiments. This unique hands-on approach ensures that the knowledge and insight that students acquire during the masterclass is action-orientated.

The programme is run every year in spring. In 2016, *International Masterclasses* were held from 11.2. to 23.3. in more than 200 research labs or universities in 46 countries (Fig. 4.3), with up to ten institutes participating per day.

Although the participating institutions are able to configure some aspects of the event individually, each masterclass follows a uniform scheme. In the morning of the day, students listen to introductory talks about particle physics. Presentations are adjusted to students' level and given by the masters—scientists familiar with outreach and education. In the lunch break, there is the opportunity for the students to talk to undergraduates, graduate students and professors.

After the lunch break, the practical part follows, which takes place on PCs. Students work with data from particle collisions recorded at the LHC on an own measurement. After an appropriate introduction to the programmes, they perform the

[1]http://www.physicsmasterclasses.org/

Fig. 4.3 Maps displaying countries that participate in *International Masterclasses*

measurement task autonomously, supported by the masters. At the end their findings are discussed with the group.

To simulate a real scientific working environment, each masterclass ends with a video conference, where up to five student groups from different countries connect

Fig. 4.4 Screenshot of a videoconference with five student groups and moderators at CERN (top left)

with two moderators at CERN or Fermilab (Batavia, Illinois, USA) to combine and discuss their results (Fig. 4.4). Students can also pick their moderators' brains in a Q&A section. Most video conferences end with a multiple-choice quiz on particle physics. More than 60 physicists have volunteered to moderate the video conferences at CERN or Fermilab.

4.3 LHC Data for High School Students

All LHC experiments provide large samples of recent data. Six measurements with a wide range of study tasks are available. For example, students can rediscover the Z boson or the structure of the proton, reconstruct "strange particles" or measure the lifetime of the D^0 particle. One of the highlights is the hunt for Higgs bosons. ATLAS and CMS have made available real Higgs candidate events for students to track this rare, elusive and very short-lived particle. Typically, a research lab selects a measurement with which the physicists have a deep relationship, guaranteeing that authentic experts are available to talk with students about what they know best. The following sections present two exemplary measurements.

The ATLAS W path[2] deals with the structure of the proton and the search for Higgs (Bilow et al. 2014). Students analyse events (Fig. 4.5) and look for a W boson decaying into a charged lepton and a neutrino (missing energy) and build the charge ratio $\frac{N_{W+}}{N_{W-}}$. The simple view of uud quarks leads to a naive approximation of $\frac{N_{W+}}{N_{W-}} = 2$.

[2]http://atlas.physicsmasterclasses.org/en/wpath.htm

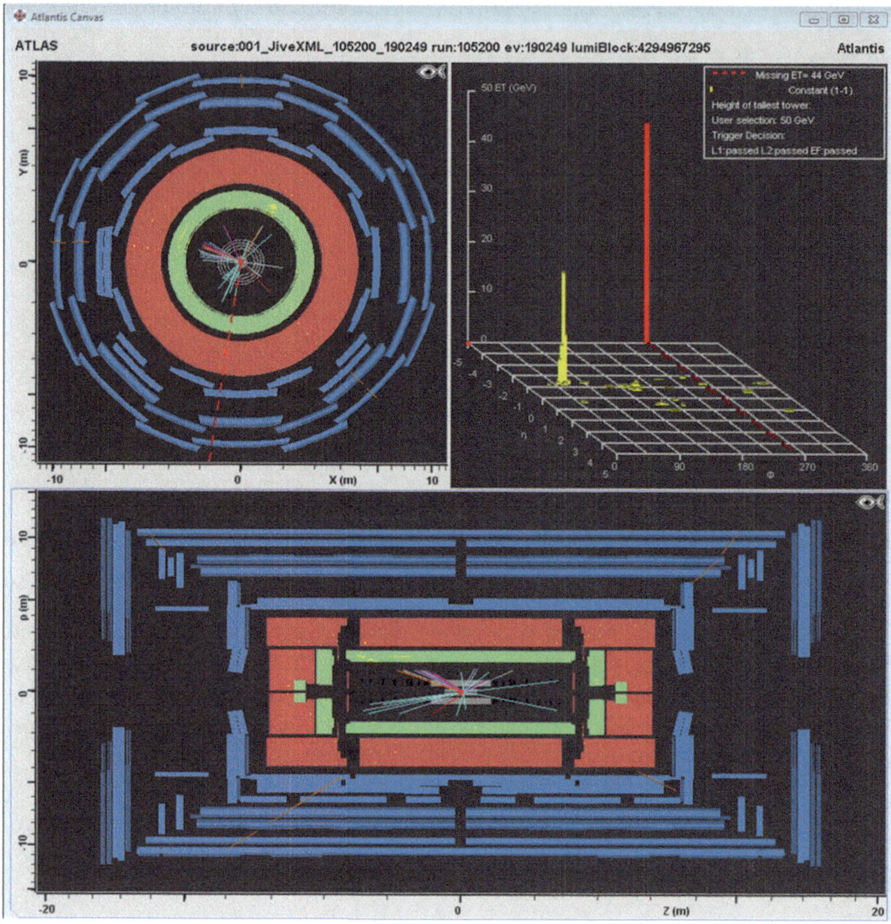

Fig. 4.5 Event display Minerva

The presence of sea quarks and gluons complicates the picture, bringing the ratio down to ~1.5, compatible with what ATLAS and CMS have measured. The next challenge is to study events containing W^+W^- pairs, characterized by two oppositely charged leptons and neutrinos. The Higgs boson decaying to W^+W^- would enhance the distribution of the azimuthal angle between the charged leptons at low values.

In the CMS measurement,[3] students use a 3D event display (iSpy-webgl). Based on the signatures of leptonic decays, they determine whether each event is a W candidate, a Z candidate, a Higgs candidate or background. For W bosons, they use the curvature of the single measurable lepton track to decide if it is a W^+ or W^- to derive the charge ratio of W boson production and characterize events as muon or

[3] http://cms.physicsmasterclasses.org/cms.html

Fig. 4.6 Students with histogram built from sticky notes

electron events to measure the electron-to-muon ratio. For Z and Higgs candidates, students put, respectively, the invariant masses of lepton and di-lepton pairs in a mass plot. They discover the Z and the Higgs peaks, including a few other resonances they might not have expected. The analysis can be done on a PC or, as shown in Fig. 4.6, by building a histogram with sticky notes.

4.4 Organization and Development of the Program

International Masterclasses are organized by the IPPOG[4] (International Particle Physics Outreach Group) collaboration (Alexopoulos et al. 2015). IPPOG is a network of scientists, science educators and communication specialists, working in informal science education and outreach for particle physics. The collaboration includes representatives from 27 states (incl. the CERN member states) and major experiments from several countries.

A steering group manages *International Masterclasses* in close cooperation with IPPOG. The coordination is provided through TU Dresden and QuarkNet: while the TU Dresden-based coordination is responsible for ~170 institutes in Europe, Africa and the Middle East, coordination through QuarkNet covers North and South America, Australia and Oceania and the Far East (~45 institutes).

International Masterclasses have developed from a programme run in the UK since 1997. The idea of particle physics masterclasses was adopted by IPPOG for the whole Europe in 2005, the World Year of Physics. At that time, masterclasses were organized in 17 countries, with 3000 students participating. With the LHC being still under construction, data from the former accelerator, CERN's LEP, was analysed. Since then the programme has experienced steady growth (Fig. 4.7). In 2006, the USA joined the programme. More countries from the Middle East or Latin America have been attracted in recent years, leading to enormous numbers of participants.

[4] http://ippog.org/

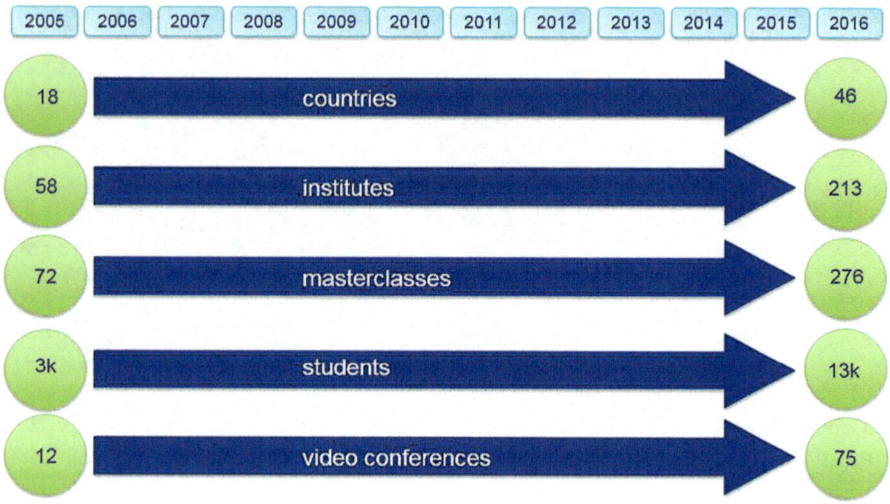

Fig. 4.7 Development of *International Masterclasses* from 2005 to 2016

4.5 The German Programme Netzwerk Teilchenwelt

International Masterclasses has led to other masterclass initiatives, e.g. national programmes, remote programmes for students in areas far from research labs, professional development programmes for teachers including masterclass elements, activities in science centres and museums and masterclasses in other fields of physics. The largest national programme exists in Germany and is called *Netzwerk Teilchenwelt*[5] (Network Particle World) (Rockstroh et al. 2013). In this multi-level programme which is led and organized at TU Dresden, 28 research labs in Germany are working in close cooperation with CERN. On its basic level, young facilitators, mostly PhD and master's students, bring CERN's data to schools and hold a masterclass. By doing so, each year about 4000 students are introduced to particle physics and get the chance to work with data from LHC experiments. The most interested students are invited to further qualification and specialization levels, which include workshops at CERN and can lead to students' own research projects (Fig. 4.8).

On parallel levels, astroparticle activities can be conducted. Students can work with detectors for cosmic rays and perform a variety of measurements, e.g. angular distribution, coincidence, determine the muon lifetime or study particle showers. Cloud chamber sets are loan to schools, and web experiments with data from the Pierre Auger observatory offer further opportunities for students to work on the subject.

[5]www.teilchenwelt.de

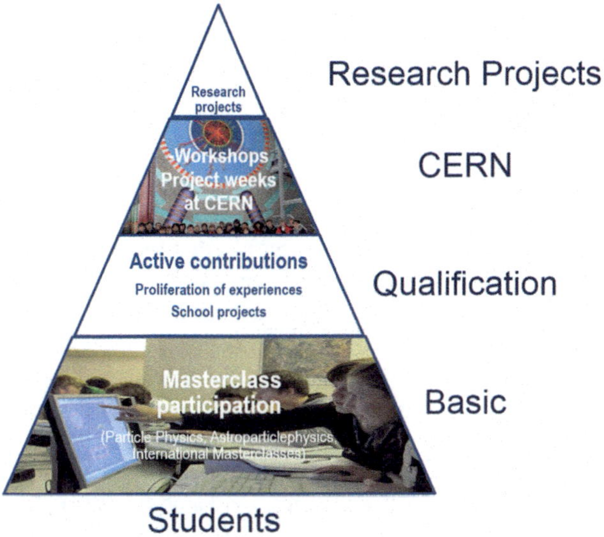

Fig. 4.8 Pyramid showing the multi-level programme of Netzwerk Teilchenwelt

Further, the wide range of activities in *Netzwerk Teilchenwelt* covers the development of various supporting material for facilitators and teachers, e.g. four volumes of teaching material for particle physics,[6] particle profiles[7] and the relaunch of the particle physics section on the largest German physics portal for schools.[8]

As described above the key feature of a masterclass is an authentic experience for high school students. They are brought as close as possible to current research and in direct contact with physicists, where they can perform own hands-on activities, using the same methods and tools like scientists. Informal learning environments like the masterclass setup are known to add valuable experiences to physics education at school and affect students' motivation and interest in science, resulting in positive effects on the students' attitudes towards physics.

4.6 Evaluation Study

The basic programme of *Netzwerk Teilchenwelt*, masterclasses with LHC data, has been evaluated concerning the issues of students' interest in particle physics and their perception of the event (Gedigk et al. 2014; Gedigk and Pospiech 2015). Different aspects of interest were studied with a questionnaire, containing items with closed answer format with a 5-point Likert scale. About 500 high school students were

[6] www.teilchenwelt.de/tp

[7] www.teilchenwelt.de/material/materialien-fuer-lehrkraefte/teilchensteckbriefe/

[8] www.leifiphysik.de/themenbereiche/teilchenphysik

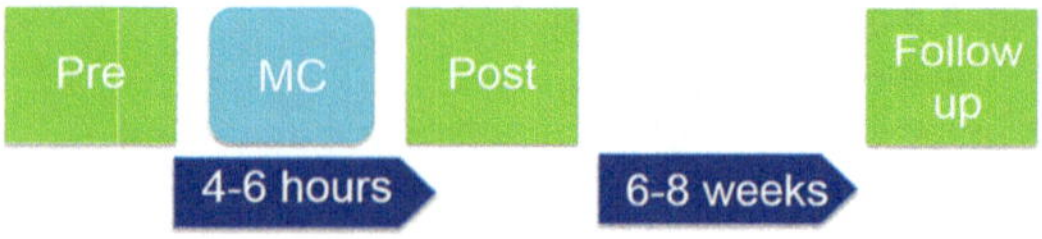

Fig. 4.9 Design of the study on students' interest and motivation before and after a masterclass (MC)

evaluated, asking questions on their interest in particle physics and the perception of the event. The study followed a pre-/post-/follow-up design (Fig. 4.9).

Results of the evaluation confirm that masterclasses are very much appreciated by students. An important feature for this success is authenticity as well as challenge. Masterclasses are especially successful for a group of students with a high interest in doing particle physics 6–8 weeks after the masterclass. This success group is remarkable large, with 26% of all participants. The investigation of this group shows the same positive effect on all students, independent on gender and level of education. Moreover, the study shows that the success group rated the features of the masterclass, e.g. challenge or authenticity, better than the other students. Students develop a long-term interest in particle physics; evidence for this is deduced from the realization of intended actions of interest in particle physics in the follow-up evaluation.

4.7 The *International Masterclasses* in Brazil

Brazil takes part in the *International Masterclasses* since 2008, starting with universities and high schools from São Paulo and Rio de Janeiro. In the following years, other cities joined the masterclasses. Figure 4.10 shows a map of Brazil indicating the cities participating in the *International Masterclasses* nowadays. It also lists the LHC experiments from which the events were analysed by each of them. Local organizers are always physicists working at universities, federal institutes or research institutes. Not all of them are members of one of the LHC experiments. Appendix lists all the local organizers and their institutions. The participants are:

- High school students and their teachers
- Teachers: physics, chemistry, geography, philosophy/sociology
- Undergraduate students, mostly from physics, but also from engineering, biblioteconomy (librarians), chemistry, philosophy

In order to participate in the videoconference with CERN and universities from other countries at the end of the activities, the local organizers in Brazil adapt the programme suggested by CERN to the time difference between Brazil and CERN in two different ways:

- Starts in the afternoon of the day before the videoconference, continues the next day, meets at noon with CERN and the other participating institutes (São Paulo, UNESP + UFABC; Curitiba, UFTPR, Rio de Janeiro, UFRJ).

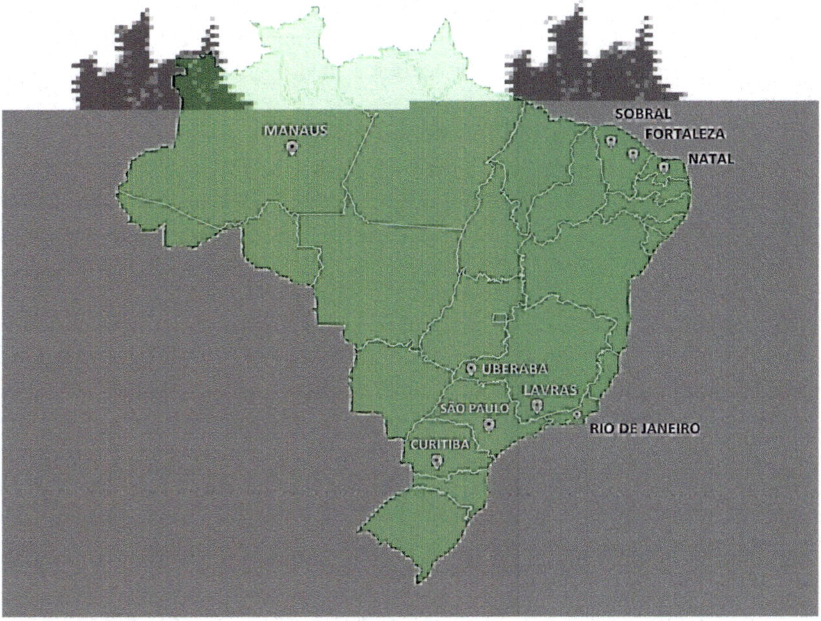

Fig. 4.10 The map of Brazil showing the cities where the *International Masterclasses* are realized, indicating from which LHC experiment are the events analysed by them

- Performs at least 3 days of activities with seminars/colloquia about particle physics, astrophysics and the experiment to be analysed in the morning and event analysis in the afternoon. The participants can also analyse some events at home and bring questions for discussions. At the day of the videoconference, final discussions are done in the morning. A virtual visit to the experiment can also be arranged, and it is normally done in the afternoon of the last day, after the meeting with CERN and other institutes (Rio de Janeiro, UERJ; Natal, UFRN + IFRN, São Paulo, USP; Lavras, UFLA; Manaus, UEAM; Uberaba, UFTM; Curitiba (UFTPR) since 2016; São Paulo, UNESP + UFABC, 2 days activities since 2013).

If the videoconference is done with Fermilab, it will happen by the end of the day. Still it follows the same organization and uses the "extra" afternoon for further discussions and the virtual visit.

The experiments ATLAS and CMS offer virtual visits allowing the students and their teachers to see the detector and the control room; depending on the time of the year, it is possible to see the detectors open. It is like being at CERN visiting the detectors. More details can be found in (http://atlasvirtualvisit.web.cern.ch/ and https://cms.cern/interact-with-cms/virtual-visits). Experiments ALICE and LHCb offer virtual tours which can be accessed at (https://www.youtube.com/watch?v=vQVEkbEvTaA, http://lhcb.web.cern.ch/lhcb/News%20of%20pit8/Pictures/LHCbVirtualTour/cern_flash.htm).

Many times the virtual visit is done before the masterclass in order to raise the interest about the experiment, LHC, CERN and particle physics. It is easier to organize, shorter and therefore easier to place in the school calendar. Of course, we talk about the *International Masterclasses*, the schools showing interest keep in contact, and the participation and collaboration begin.

Since 2008 different ways to reach the students and their teachers have been used. The best, so far, are personal contact with the schools and the teachers, Facebook pages (Masterclass BR, Masterclass-RN, MasterClass RN, Masterclass UFTM, LAPE UFRJ, SPRACE, LHC–CP2), web pages (https://www.sprace.org.br/masterclass), posters sent to the schools (Fig. 4.11) and email. The less effective are press releases: short texts in newspapers, magazines and internal journals. Twitter has worked only with UFRJ, in Rio de Janeiro. Orkut (long gone), Google Groups and similars showed no effectiveness at all.

The Brazilian Physical Society organizes a National Program for Physics Teachers together with the Portugal-CERN School, in Portuguese, where teachers from Brazil can take part through the CERN School of Physics Program. It is coordinated by Prof. Nilson Garcia from UFTPR. Some of those teachers contact

Fig. 4.11 Example of the posters sent to Brazilian high schools to announce the *International Masterclass*

us once they are back from CERN and join the group participating in the *International Masterclass* later on.

Beyond the masterclasses in March–April, the time of the videoconferences, other activities are organized:

- São Paulo (USP) offers a course about nuclear and particle physics to (mostly) physics teachers during the USP-Escola (USP-School) (http://hepic.if.usp.br/?q=pt-br/evento/curso-de-extensão-inserção-da-f%C3%ADsica-moderna-no-ensino-médio-através-do-estudo-dos).
- São Paulo (USP) and Rio de Janeiro (UFRJ and UERJ) do the masterclasses-hands-on particle physics along the year in different schools, as long as the teachers show interest in doing it.
- São Paulo (UNESP/SPRACE) realized in 2012 and 2015 a workshop for physics teachers and undergraduate physics students who will become teachers.
- The LHCb group at UFRJ (Rio de Janeiro) is developing a new exercise for the LHCb Masterclass.
- To help the learning process, São Paulo (UNESP/SPRACE) and Rio de Janeiro (UERJ) develop educational games.
- To remember the masterclass, Rio de Janeiro (UERJ) and Natal (UFRN+IFRN) teach the students to produce mugs (Fig. 4.12) or T-shirts. It is also a good publicity.
- Local organizers of the masterclass in Brazil normally work together with physicists from physics education, producing publications, master's and PhD theses in this area. The group at the University of São Paulo (USP) is the most successful in that topic.

The participants of the *International Masterclasses* are not followed after it is finished, but many of them get in contact with us. Most of them did the masterclasses in a 3-day period. All of them continue to study, not necessarily physics. Some participants become undergraduate research fellows under our supervision and later on, even master's and PhD students, continue to work with us.

Fig. 4.12 Example of mugs produced by the masterclass students in Natal (UFRN) and Rio de Janeiro (UERJ)

The participants may give a feedback if they want to. It is not mandatory. It can be anonymous (written in a paper) or not (through email, WhatsApp, Facebook, even spoken). It helps to improve the activities, the plan for next year. They can be resumed as:

- Students ask to participate again in the Masterclass, next year.
- Most common complaint: “It is too short!”
- “It is hard work!” (not as a complaint).
- From the teachers: It helps in the process to bring modern and present physics to the students at high school level. Although everything is new, it is not a burden; on the other way round, it awakens the interest of the students.
- From the geography teacher: It shows the students there is a big world out there.
- From the students: We learn to work in group, to help each other and everything gets better.
- From the students: We did not know in time about this activity. Why don’t you put it in WhatsApp and all social media?
- About the mugs: “Everybody loves it!”
- Drawing the events in a paper using different pens or painting the events: “Never do it again. It is too boring!” Therefore this activity was cancelled.
- Best comment I (Marcia Begalli) ever received: “You will not dare to stop this amazing activity right now, right?” just after the video conference with CERN (March, 23, 2016).

4.8 Conclusion

The format of masterclasses which is used in both programmes, *International Masterclasses* and *Netzwerk Teilchenwelt*, as well as the ones used in Brazil, has proven to affect students’ attitude towards physics, as well as in science and technology in general, and to be able to sustain interest independently from gender and class or school form. The programmes have an inherent potential to inspire high school students for modern physics. Under this aspect, the further growth of national and international programmes using the format of masterclasses and their spread to more and more parts of the world are highly beneficial.

Appendix

The Masterclass Local Organizers in Brazil

Rio de Janeiro, RJ

Marcia Begalli, Vitor Oguri
UERJ (State University of Rio de Janeiro)

Miriam Gandelman, Murilo Rangel, Irina Nasteva
UFRJ (Federal University of Rio de Janeiro)

São Paulo, SP

Sandra Padula, Valeria Dias, Fernando L. Campos Carvalho, Nelson Barrelo, Cleide Rizzatto
UNESP/SPRACE (University of the State of São Paulo)
Pedro Mercadante, Giselle Watanabe, Eduardo Gregores, Lucio Costa
UFABC (Federal University of [the] ABC [region])
Marcelo Munhoz, Ivã Gurgel, Graciella Watanabe
USP (University of São Paulo)

Curitiba, PR

Nilson Garcia, Luciana Rocha Hirsch
UTFPR (Federal Technical University of Paraná)

Lavras, MG

Luiz Cleber Tavares de Brito, Helvécio Fargnoli
UFLA (Federal University of Lavras)

Uberaba, MG

Marcos Dionízio Moreira, Álvaro Gomes dos Santos Neto
UFTM (Federal University of [the] Triângulo Mineiro [region])

Natal, RN

Anderson Guedes, Ronai Lisboa
UFRN (Federal University of Rio Grande do Norte)
Amadeo Albino Jr.
IFRN (Federal Institute of Rio Grande do Norte)
Marcia Begalli (UERJ)

Manaus, AM

Alberto Santoro, Luciana Cunha
UEAM (University of the State of Amazonas)
Marcia Begalli (UERJ)

Fortaleza, CE

Mairton Cavalcante Romeu
IFCE (Federal Institute of Ceará)
Marcia Begalli (UERJ)

Near Future: (Rio de Janeiro, RJ) André Massaferri, André Morais
CBPF (Brazilian Center for Physics Research)

References

Alexopoulos, A., Barney, D., Bilow, U., Adam Boudarios, C., Kobel, M., Kourkoumelis, C., Melo, I., Rangel Smith, C. (2015): Resources for Education and Outreach Activities: Discussion Session. In: Proceedings of Science EPS-HEP (2015) 619. https://inspirehep.net/record/1430777/files/PoS(EPS-HEP2015)619.pdf

Bardeen, M., Beck, H., Bilow, U., Cecire, K., Ould-Saada, F., Kobel, M. (2014): International Masterclasses in the LHC era. CERN Courier 6 (2014). http://cerncourier.com/cws/article/cern/57305

Bilow, U., Kobel, M. (2014): International Masterclasses – bringing LHC data to school children. EPJ Web Conf. 71 (2014) 00018. https://doi.org/10.1051/epjconf/2014-71000-18

Bilow, U., Hasterok, C., Jende, K., Kobel, M., Rudolph, C., Woithe, J. (2014): ATLAS W path – real data from the LHC for high school students. EPJ Web Conf. 71 (2014) 00017 https://doi.org/10.1051/epjconf/2014-71000-17

Gedigk, K., Pospiech, G. (2015): Development of students' interest in particle physics as effect of participating in a Masterclass. Il Nuovo Cimento 38 C (2015) 100. https://doi.org/10.1393/ncc/i2015-15085-2

Gedigk, K., Kobel, M., Pospiech, G. (2014): Development of interest in particle physics as an effect of school events in an authentic setting. In: Dvorak, L.: ICPE-EPEC 2013 Conference Proceedings (pp. 396–404). http://www.icpe2013.org/uploads/ICPEEPEC_2013_ConferenceProceedings.pdf

Rockstroh, M., Bilow, U., Gedigk, K., Glueck, A., Kobel, M., Pospiech, G. (2013): Netzwerk Teilchenwelt – Hands On Particle Physics Masterclasses in Germany – Best Practice in Sharing Authentic Science with the Public. In: Lazoudis, A.: Discover the Cosmos Conference Proceedings (pp. 61–70). http://handsonuniverse.org/wp-content/uploads/2013/07/DtC_conf_proceedings.pdf

Part II
Evaluations About Established Methods of Teaching Physics

Chapter 5
A Mathematical Model of Peer Instruction and Its Applications

Hideo Nitta

Abstract In this chapter, we review mathematical models of learning. The focus is on the mathematical model of peer instruction (PI) based on the master equation that describes the dynamic change of students' response in PI. In this model, for evaluating the effectiveness of each question for PI, the "peer instruction efficiency (PIE)" is introduced in analogy with the Hake gain. It is shown that, in the simplest approximation, PIE becomes proportional to the relative number of students answering correctly before discussion. The mathematical model is applied to introductory physics courses at a university and a high school. It is found that overall practical data of PIE moderately agree with theory. Application of theoretical results to practical data, such as identifying effective PI questions, is also discussed.

5.1 Introduction

Physics describes the properties of physical substances and interaction among them. Then, human beings as "physical substances" composed of atoms and molecules should, in principle, be described by physics. Of course, at the present stage of physics, it is absolutely a desperate attempt to construct a mathematical theory of dynamics, including learning, for a student. However, it may become possible to describe a learning process for *many students* if we express it with only a few "macroscopic variables." This describing is analogous to statistic-mechanical explanation of macroscopic properties of a gas. Although it is eventually impossible to predict the motion of molecules of the gas, the thermodynamic variables, such as pressure, specific heat, etc., are very well described by theory.

In this chapter, we first introduce a few mathematical models of learning that developed previously. Then we present our mathematical model that describes dynamics of the response of students in peer instruction (PI). In this model, for evaluating the effectiveness of each question for PI, the "Peer-Instruction Efficiency

H. Nitta (✉)
Department of Physics, Tokyo Gakugei University, Tokyo, Japan
e-mail: nitta@u-gakugei.ac.jp

M. Pietrocola (ed.), *Upgrading Physics Education to Meet the Needs of Society*,
https://doi.org/10.1007/978-3-319-96163-7_5

(PIE)" is introduced in analogy with the Hake gain. It is shown that, in the simplest approximation, PIE becomes proportional to the relative number of students answering correctly before discussion. The mathematical model is applied to introductory physics courses at a university and physics classes at a high school. It is found that overall practical data of PIE moderately agrees with theory. Application of PIE to data analysis is also discussed.

5.2 The Hake Gain

The recent trend in physics education research is to evaluate the effect of teaching in a quantitative manner. This trend originated from the celebrated work by Hake (1998), in which the normalized learning gain (the Hake gain) was introduced for evaluating the students' achievement in the class. The Hake gain is defined by

$$\langle g\rangle = \frac{\langle S_{\mathrm{post}}\rangle - \langle S_{\mathrm{pre}}\rangle}{100 - \langle S_{\mathrm{pre}}\rangle}, \tag{5.1}$$

where $\langle S_{\mathrm{pre}}\rangle$ and $\langle S_{\mathrm{post}}\rangle$ are the class-average scores of the pretest and posttest, respectively. The Hake gain makes numerical comparison of the effectiveness of various teaching methods possible. By using Eq. (5.1), Hake showed that the average value of $\langle g\rangle$ for the interactive engagement classes is more than twice of $\langle g\rangle$ for the traditional lectures. This surprising result not only triggered the worldwide spread of the active learning method in physics classes but also impressed researchers into the importance of quantitative evaluation of teaching methods.

5.3 A Generalized Ising Model of Teaching-Learning Process

Although there have been a lot of interests in evaluating teaching methods quantitatively, only a few mathematical theories of teaching-learning process have been developed. An interesting theory was developed by Bordogna and Albano (2001, 2003) to simulate teaching-learning process using the Monte-Carlo method. Their model, which we call the "BA model", is based on a generalized Ising model that has been used for spin systems, neural networks, and social systems. The BA model assumes that the knowledge of the jth student at time t is given by $\sigma_j(t)$ which satisfies $-1 \leq \sigma_j(t) \leq 1$, where $\sigma_j(t) = 1$ represents perfect knowledge of the target subject considered. The knowledge $\sigma_j(t)$ develops by the effects of "cognitive impact (CI)" acting on the student. The BA model considers three types of CI: $CI^{\mathrm{TS}}(j,t)$, $CI^{\mathrm{SS}}(j,t)$, and $CI^{\mathrm{BS}}(j,t)$. $CI^{\mathrm{TS}}(j,t)$ represents the CI of the teacher on the jth student at time t, $CI^{\mathrm{BS}}(j,t)$ the student-student interaction, and $CI^{\mathrm{BS}}(j,t)$ the bibliography and other sources of information. By performing the Monte-Carlo simulation, Bordogna and Albano showed that the learning achievements become higher for students engaging in collaborating work than for those only attending lectures. This result is consistent with finding by Hake mentioned above. They also showed that the structure of students' groups may influence the achievements. The result of simulation indicated that lower-knowledge students may learn at the expense of their higher-knowledge peers.

5.4 Mathematical Learning Models by Pritchard Et al.

Pritchard et al. developed mathematical models that describe learning processes of tabula rasa, constructivism, and tutoring (Pritchard et al. 2008). Their aim of developing the theory is to provide a quantitative tool that allows the parametrization of measured learning data through the Hake gain, Eq. (5.1). They assumed that a concept inventory (CI) covers the knowledge domain T composed of the known knowledge domain $K_T(t)$ and the unknown knowledge domain $U_T(t)$. The knowledge domain is normalized to satisfy the relation $K_T(t)+U_T(t) = 1$, where t is the amount of teaching or instruction.

In the pure memory model, Pritchard et al. assumed the following simple differential equation:

$$\frac{dU_T(t)}{dt} = -\alpha_m U_T(t), \tag{5.2}$$

where α_m represents the sticking probability of taught things in student's mind for the pure memory model.

In the simple connected model, based on the idea of constructivism that learning occurs by constructing an association between the new knowledge and prior knowledge, Pritchard et al. assumed the logistic differential equation

$$\frac{dU_T(t)}{dt} = -\alpha_c U_T(t) K_T(t), \tag{5.3}$$

which can be solved analytically. They also proposed the combined model of pure memory and constructivism called the connectedness model.

In the tutoring model, it is assumed that knowledge grows in a uniform rate of learning:

$$K_T(t) = \alpha_{tu} t + K_T(0), \tag{5.4}$$

where $K_T(0)$ represents the initial knowledge before tutoring.

Based on the analytic solutions of these differential equations, Pritchard et al. obtained an analytic expression of the Hake gain for each learning model. They compared the analytic expressions with data and found reasonable agreement (Pritchard et al. 2008).

5.5 Mathematical Theory of Peer Instruction

Peer instruction (PI) is one of the simplest methods of instruction for increasing students' engagement in lectures (Mazur 1997). In a PI-based lecture, the teacher presents a concept-oriented multiple-choice question (MCQ) about the material just covered by his/her lecture. Then students are encouraged to discuss the question with their neighbor. Due to its simplicity as well as having no limitations on the class size, PI is suitable for introducing essence of active learning to an old-fashioned lecture without changing the curriculum.

For the success of a PI-based lecture, it is crucial to present an effective MCQ for discussion. A good MCQ should induce intensive discussion among students as well as effectively support students to understand the target concept. For the purpose to evaluate MCQs, it is important to develop a measure of their effectiveness. To obtain such a measure, we try to express the dynamics of peer instruction by a mathematical model.

We consider the following process. First, students in a class are posed a MCQ. Each student considers the MCQ without discussion to give her/his own answer. Then students discuss the MCQ with their neighbor, exchanging ideas about their answers.

Let us first consider the dynamics describing the change of the number of students who answer correctly before and after peer discussion for a MCQ. We define that $\rho_{\rm b}(q, a)$ and $\rho_{\rm a}(q, a)$ are the normalized numbers of students choosing the answer a for the MCQ (denoted by q) before and after discussion, respectively, and that $T_{a'a}(q)$ is the transition rate from an answer a to the other one a' by peer discussion. Then $\rho_i(q, a)$ (i = a,b) satisfy the following master equation (Nitta 2010):

$$\rho_a(q,c) - \rho_{\rm b}(q,c) = -\sum_{d(\neq c)} T_{dc}(q)\,\rho_{\rm b}(q,c) + \sum_{d(\neq c)} T_{cd}(q)\rho_{\rm b}(q,d) \tag{5.5}$$

where c and d represent the correct answer and the incorrect answers, respectively. The left-hand-side of Eq. (5.5) represents the difference of the number of students answering correctly before and after discussion. The first term on the right-hand side (r.h.s.) of Eq. (5.5) represents the normalized number of students who at first choose the correct answer and then, after discussion, change it into one of the incorrect answers. This process is usually called the "outgoing process" in the theory of irreversible statistical mechanics. Similarly, the second term on the r.h.s. of Eq. (5.5) represents the normalized number of students who give incorrect answers before discussion and the correct answer after discussion (i.e., the "incoming process"). Of course, $\rho_a(q, c)$, $T_{dc}(q)$, etc. are extremely simplified functions. We have neglected enormous number of parameters such as the quality of MCQ; students' character, knowledge, and reasoning ability; influence of teachers; literature; quality of peer discussion; and many others. It should be noted that $T_{dc}(q) \neq T_{cd}(q)$, i.e., the detailed balance which is normally assumed in physics problems does not hold here. Indeed, if PI is an effective education method at all, it should result in $T_{dc}(q) \ll T_{cd}(q)$.

One may feel that Eq. (5.5) is rather hypothetical. However, this is not so. We emphasize that Eq. (5.5) is, though phenomenological, an exact equation in that all variables and functions in Eq. (5.5) can be determined by PI data for a MCQ. Indeed, by using an audience response system, called "clicker system," the teacher can collect all responses from students for the MCQ, q, before and after the discussion session. Then, from the response data, one can determine the transition probability, $T_{a'a}(q)$, as well as the normalized number of students, $\rho_{\rm b}(q, a)$ and $\rho_{\rm a}(q, a)$.

Although Eq. (5.5) is exact in the above sense, it cannot predict anything without providing a certain analytical expression for $T_{a'a}(q)$. Naturally, as mentioned before, it is not realistic to derive such an expression from the first principle because it should depend on enormous number of student parameters. Here we take a phenomenological approach for obtaining an analytical expression.

Let us neglect

1. The transition from the correct answer to an incorrect answer, i.e., the first term of the r.h.s. of Eq. (5.5)
2. The dependence of $T_{cd}(q)$ on incorrect answers
3. The MCQ dependence, q

Then we obtain,

$$\rho_{\rm a}(c) - \rho_{\rm b}(c) = T \sum_{d(\neq c)} \rho_{\rm b}(d), \tag{5.6}$$

where T stands for the "average" transition rate from incorrect answers to the correct answer. Using the identity $\sum_{d(\neq c)} \rho_{\rm b}(d) = 1 - \rho_{\rm b}(c)$, which means that "the (*normalized*) number of students giving the incorrect answers" is equal to "the number of all students" minus "the number of students giving the correct answer," we obtain

$$\rho_{\rm a} - \rho_{\rm b} = T(1 - \rho_{\rm b}), \tag{5.7}$$

where now $\rho_{\rm b}$ and $\rho_{\rm a}$ represent simply the normalized number of students giving the correct answer before and after discussion, respectively. At this stage, only one parameter, T, i.e., the transition rate from incorrect answers to the correct answer, is left. Here we further assume that T is the function of $\rho_{\rm b}$: $T = T(\rho_{\rm b})$. Then we may expand T into the power series:

$$T = k_0 + k_1 \rho_{\rm b} + k_2 \rho_b^2 + \cdots, \tag{5.8}$$

where k_i ($i = 0, 1, 2, \cdots$) are constants. Under the condition that $\rho_{\rm b} = 1 \rightarrow T = 1$ and $\rho_{\rm b} = 0 \rightarrow \rho_{\rm a} = 0$, we have $k_1 = 1$ and $k_0 = 0$; hence $T = \rho_{\rm b}$. This simple result suggests that the more the number of students answering correctly before discussion increases, the more transition from incorrect answers to the correct answer happens. Using $T = \rho_b$ with Eq. (5.7), we obtain

$$\rho_a - \rho_b = \rho_b(1 - \rho_b), \tag{5.9}$$

For evaluating the effectiveness of PI, let us introduce the "Peer Instruction Efficiency" (PIE) for a MCQ q by

$$\eta(q) = \frac{\rho_a(q,c) - \rho_b(q,c)}{1 - \rho_b(q,c)}. \tag{5.10}$$

Although the definition of PIE looks the same as the Hake gain, they are different in character. The Hake gain represents the overall learning gain for a series of lectures as a whole, whereas PIE represents the efficiency of students' discussion for each MCQ.

Substituting Eq. (5.9) into Eq. (5.10), we obtain the very simple expression for PIE:

$$\eta = \rho_b. \tag{5.11}$$

We will discuss the use of PIE in the next section.

Let us generalize the idea of using the master equation for learning processes. Instead of $\rho_{a,b}(q,c)$, we consider the fraction of students in the correct stage of knowledge c about a concept q at time t, $p(q;c;t)$. Following the idea of Eq. (5.5), we assume that the development of knowledge during the time Δt is described by the master equation

$$\Delta p(q;c;t) = -\sum_{d(\neq c)} T_{dc}(q;t)p(q;c;t) + \sum_{d(\neq c)} T_{cd}(q;t)p(q;d;t), \tag{5.12}$$

where $\Delta p(q;c;t) = p(q;c;t+\Delta t) - p(q;c;t)$ and d represent incorrect or blank state of knowledge about the concept q and $T_{cd}(q;t)$ the transition rate of the state of knowledge from d to c during the time interval Δt. If we neglect the "outgoing process" of the knowledge construction, i.e., the first term of the r.h.s. of Eq. (5.12), and make other similar approximations as before, we obtain

$$\Delta p(t) = (\alpha \Delta t)(1 - p(t)) \tag{5.13}$$

where we have omitted the parameters q and c and denoted the transition rate to the correct state of knowledge per unit time as α. Further, we neglect the explicit time dependence of α but assume that α is the function of $p(t)$. Then, by expanding α into the power series and taking the lowest-order terms like Eq. (5.8), we obtain

$$\alpha = k_0 + k_1 p(t), \tag{5.14}$$

where k_0 and k_1 are constants. Substituting Eq. (5.14) into Eq. (5.13) and taking the limit $\Delta t \to 0$, we obtain the differential equation

$$\frac{dp(t)}{dt} = (k_0 + k_1 p(t))(1 - p(t)), \tag{5.15}$$

which is essentially the same differential equation given by Pritchard et al. for their "connectedness model" (Pritchard et al. 2008). It is worthwhile to note that if we set $k_0 = \alpha_m$, $k_1 = 0$, and $1 - p(t) = U_T(t)$, Eq. (5.15) is reduced to the pure memory model of Eq. (5.2). On the other hand, by setting $k_0 = 0$, $k_1 = \alpha_c$, and $p(t) = K_T(t)$, we recover the simple connected model of Eq. (5.3).

5.6 Applications

Now we compare our theoretical result with class data. In Fig. 5.1, we compare the theoretical curve given by Eq. (5.9) and data of the fraction of correct answer before and after peer discussion. The data was taken from an introductory physics course for the first year students in TGU during the academic years 2009–2011 (Nitta et al.

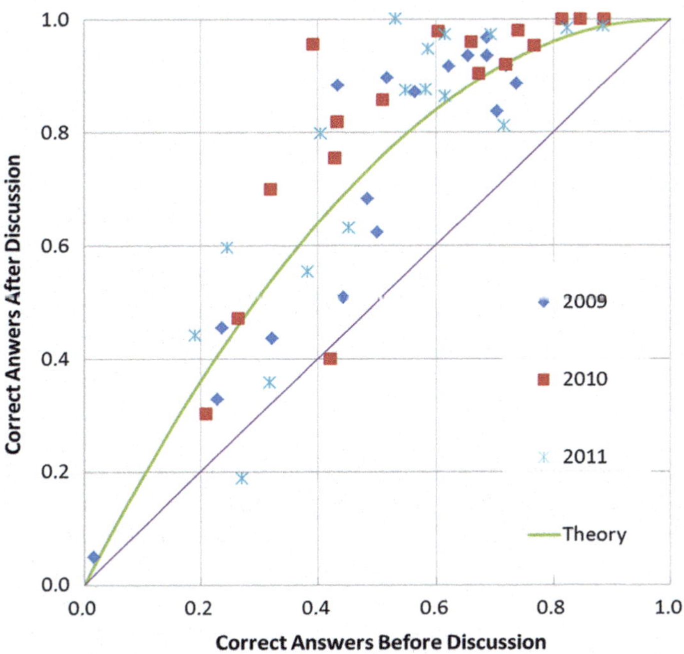

Fig. 5.1 Comparison of theory with data of the fraction of correct answer before and after peer discussion (Nitta et al. 2014)

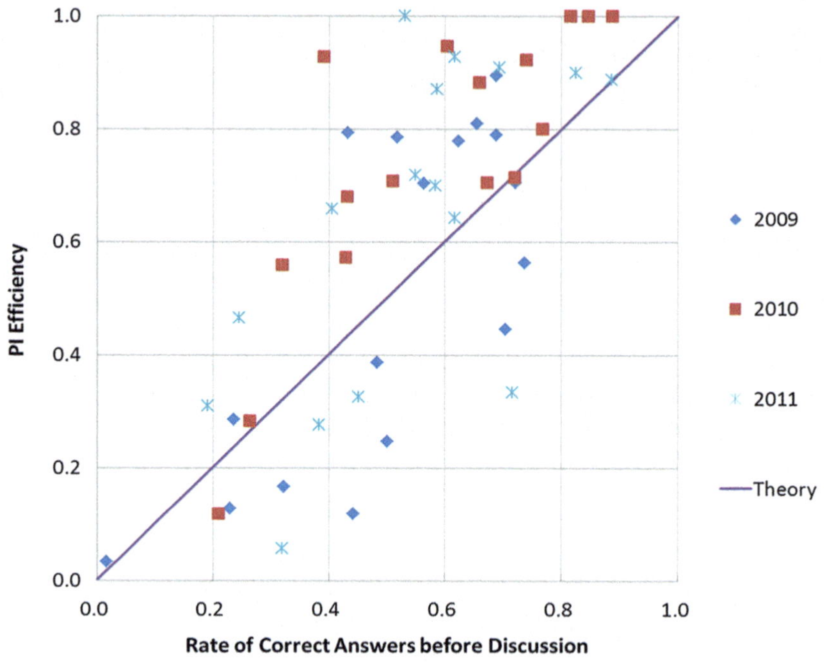

Fig. 5.2 The peer instruction efficiency

2014). The numbers of enrolled students were 60 (2009), 55 (2010), and 81 (2011). Teacher A taught 2009 class while teacher B taught 2010 and 2011 classes. The contents of the courses and MCQs were almost the same between teacher A and teacher B. In Fig. 5.1, data from 17 MCQs in each year were plotted, i.e., 51 plots altogether are shown.

In Fig. 5.2, the same data of Fig. 5.1 are represented in the form of PIE. The straight line corresponds to the simple theoretical expression of PIE, $\eta = \rho_b$. Although the data are dispersed, the theoretical line roughly agrees with the data. This agreement indicates that the approximation $T = \rho_b$ is basically valid. In other words, the transition rate from the incorrect answers to the correct answer by peer discussion tends to increase as the number of students answering correctly before discussion increases.

In Figs. 5.3 and 5.4, we show the data of PI taken from an introductory physics course in TGU High School. Only the data about mechanics are shown. The number of plots is 161, which have been gathered from 2008 to 2016. Each plot represents responses of about 40–120 students for 1–3 classrooms. The best-fit curves in Figs. 5.3 and 5.4 are given by, respectively, $y = -1.2x^2 + 2.3x - 0.069$ and $y = 1.0x - 0.012$. In Fig. 5.4, although the best-fit line agrees well with the theoretical line, data seem to deviate from theory at the region where the rate of correct answers before discussion is low, typically ρ_b is less than 0.2. In this region, almost every point is lower than theoretical line. This indicates that discussion is

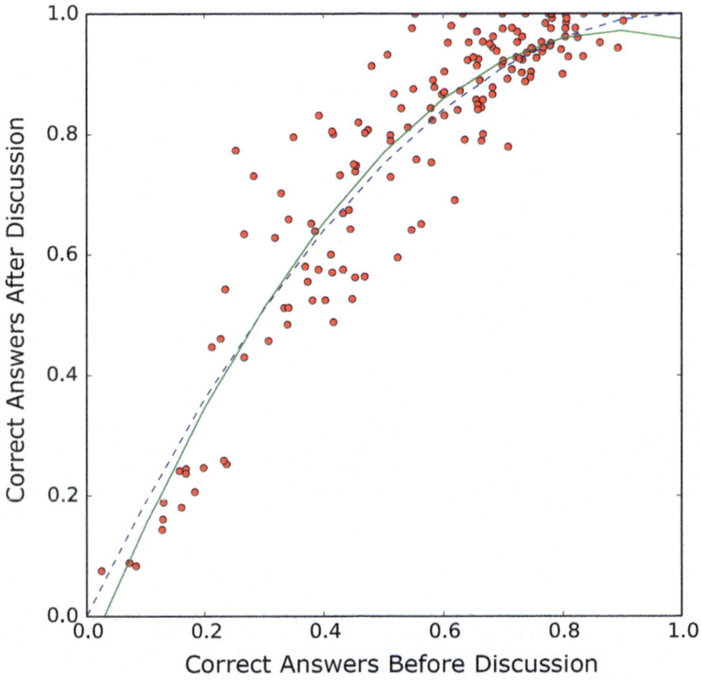

Fig. 5.3 Comparison of theory with data of the fraction of correct answer before and after PI for high school classes. The dashed line and the solid line represent the theoretical curve and the best-fit curve in the quadratic function, respectively

ineffective when there is not enough fraction of students who can lead the discussion into the right direction.

Our aim to introduce PIE has been for evaluating the effectiveness of each PI. We shall try to demonstrate such usage of PIE for the evaluation of PI or MCQs. Figure 5.5 represents the deviation of the PIE data, shown in Fig. 5.2, from the theoretical value. The deviation δ is defined by

$$\delta = \mathrm{PIE(datum)} - \rho_b, \tag{5.16}$$

The average value and the standard deviation of δ are $\overline{\delta}_{\mathrm{tot}} = 0.062$ and $\sigma_{\mathrm{tot}} = 0.22$, respectively. These values can be used for evaluating the effectiveness of MCQs grouped in specific subjects. For example, let us consider the δ for MCQs about interpretation of kinematics graphs. The frequency distribution of δ for MCQs about interpretation of kinematics graphs is shown in Fig. 5.6. By the one-tailed Welch's t-test, the average value of delta, $\overline{\delta}_{\mathrm{graph}} = 0.20$, turns out to be larger than $\overline{\delta}_{\mathrm{tot}}$ with the significance level of $p < 0.05$. In other words, PI for interpretation of kinematic graphs is more effective than other types of PI. We may interpret this result in the following way. The difficulties students have on understanding kinematic graphs are

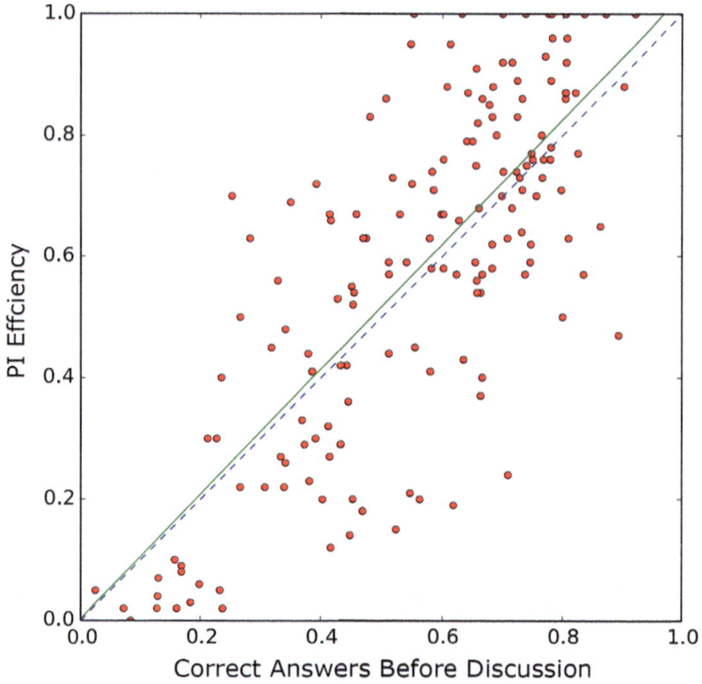

Fig. 5.4 PIE for high school classes. The dashed line and solid line represent the theoretical line and the best-fit line by the linear function, respectively

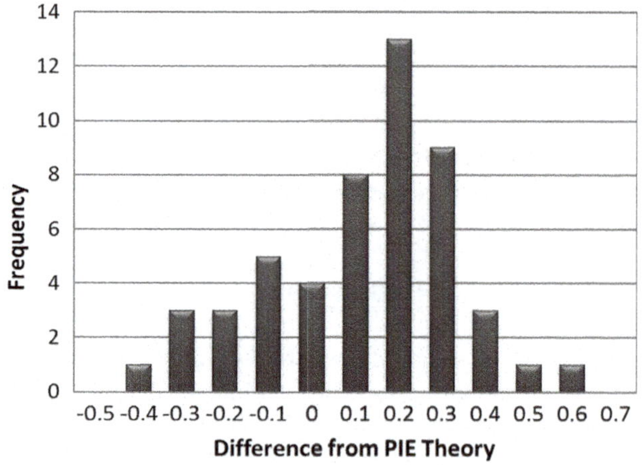

Fig. 5.5 Deviation of PIE data from theory

not conceptual difficulties but rather simple technical problems that can be easily resolved by instruction from other students who have already overcome the technical difficulties. This result indicates that interpretation of kinematic graphs is one of the best subjects for PI in that students' difficulties can be overcome very effectively by

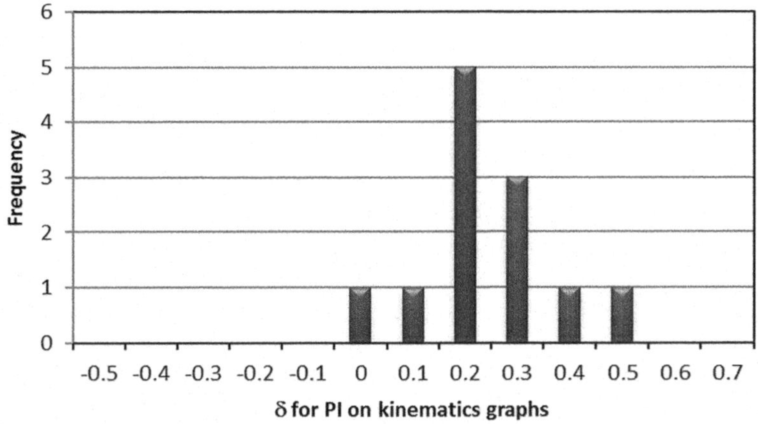

Fig. 5.6 The distribution of δ for PI on interpretation of kinematics graphs

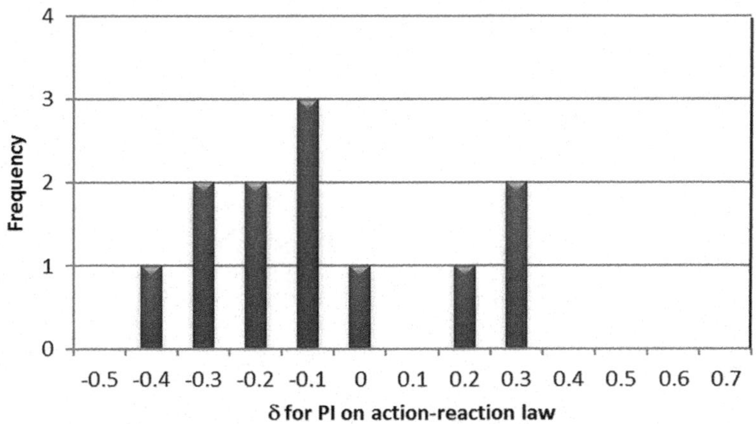

Fig. 5.7 The distribution of δ for PI on action-reaction law

instruction among students. In other words, on this specific subject, instruction by peers will be more effective than that by a teacher. It should be noted, however, that interpretation of graphs on kinematics are rather special than other subjects related to laws of physics, such as graphs on force and motion. Since the relation among position, velocity, and acceleration is mathematical, the origin of students' difficulties would not come from understanding of the law of physics but from "technical difficulties" on interpreting mathematical meaning of graphs.

Another example of the distribution of δ is shown in Fig. 5.7 for MCQs about action-reaction law. In this case, the average value of delta, $\overline{\delta}_{\mathrm{ar}} = -0.13$, is significantly lower than $\overline{\delta}_{\mathrm{tot}}$ ($p < 0.05$). This result implies that students have so robust naïve conceptions on action-reaction forces that they do not likely to change their beliefs by peer discussion.

5.7 Concluding Remarks

In this chapter, we have presented a mathematical theory of peer instruction (PI) that describes the dynamics of student discussion. The derived simple expression gives a kind of standard line that reasonably agrees with data. We have demonstrated that PIE is useful for rough estimations of the effectiveness of multiple-choice questions (MCQs) for students' discussion. If PIE is larger than the normalized rate of correct answers before discussion, then the MCQ used for the PI turns out to be, roughly speaking, more effective than standard; if smaller, less effective. Although this evaluation is very rough, we find it useful for improving MCQs and lectures in our practice for years.

Finally, we would like to point out that, by combining data of PIE with the results of pre/post concept inventory, such as FCI, one may obtain rich information about students' naïve conceptions as well as the effectiveness of her/his PI-based lectures (Nitta et al. 2014).

The author would like to thank Tomoshige Kudo and Takuya Aiba for useful discussions and their help of data analysis as well as many graduate students who taught introductory physics classes in TGU High School for getting the data of PI.

References

C. M. Bordogna and E. V. Albano, Theoretical Description of Teaching-Learning Processes: A Multidisciplinary Approach Phys. Rev. Lett. **87**, 118701 (2001).

C. M. Bordogna and E. V. Albano, Simulation of social processes: application to social learning Physica A **329**, 281 (2003).

R. R. Hake, Interactive-engagement versus traditional methods: a six-thousand-student survey of mechanics test data for introductory physics courses Am. J. Phys. **66**, 64 (1998).

E. Mazur Peer instruction, a user's manual (Pearson-Prentice Hall, New Jersey, 1997).

H. Nitta, Mathematical theory of peer-instruction dynamics, Phys. Rev. Spec. Top. Phys. Educ. Res. **6**, 020105 (2010).

H. Nitta, S. Matsuura, and T. Kudo, Implementation and analysis of peer-instruction in introductory physics lectures, J. Sci. Educ. Japan, **38**, 12 (2014) (*in Japanese*).

D. H. Pritchard, Y-J. Lee, and L. Bao, Mathematical learning models that depend on prior knowledge and instructional strategies, Phys. Rev. Spec. Top. Phys. Educ. Res. **4**, 010109 (2008).

Chapter 6
Simple Experiments in Physics Teaching and Learning: Do They Have Any Perspectives?

Leos Dvorak

Abstract Being cheap, requiring just a short time and often providing interesting results, simple experiments have been used in physics classes for a long time. However, can they compete with what nowadays ICT, modern technologies and the Internet offer to us and our students in more and more attractive forms and formerly unimaginable range and quality? Or are they rather obsolete and, metaphorically speaking, "endangered species"? The purpose of the chapter is to defend the opinion that "simple experiments are not dead". To do it, we try to look at them from a more general point of view offering sometimes maybe unorthodox but hopefully interesting views and perspectives. We claim and demonstrate that simple experiments are "not stupid" (i.e. they are challenging at many levels, enable even quantitative measurements, etc.), adaptable (here, we borrow, rather as metaphors, some ideas from evolution biology), multipurpose tools that can teach us a lot (not only facts) and, finally, though they can serve many purposes, they are also valuable just by themselves. These general ideas and views are supplemented and illustrated by examples of series of simple experiments from various parts of physics: electrostatics, charges and currents, sound and waves, magnetism and mechanics. About twenty experiments are presented, some of them hopefully in new or not so known variants. From all these considerations and examples we conclude that simple experiments will stay with us and we will use and enjoy them in physics education even in times to come.

6.1 Introduction

The role of experiments (or, more generally, the role of practical work) in physics education has been a topic of many research studies, papers, essays and reviews. The importance of experiments was stressed in many projects, from large and famous as PSSS or Nuffield to a lot of more local ones. In recent years, inquiry-based science

L. Dvorak (✉)
Faculty of Mathematics and Physics, Department of Physics Education, Charles University, Prague, Czech Republic
e-mail: Leos.Dvorak@mff.cuni.cz

M. Pietrocola (ed.), *Upgrading Physics Education to Meet the Needs of Society*,
https://doi.org/10.1007/978-3-319-96163-7_6

education is a term used so frequently that it seems nearly impossible to avoid it in projects aimed at improving physics education. Therefore, one can think that we live in a "golden age" of experiments when nothing can question their fundamental role for physics teaching and learning and their future can be even brighter. In other words, is it not true that every physics teacher and educator swears on the well-known mantra *"I hear it, I forget, I see it, I remember, I do it, I understand"*?

However, many studies warn us that things are not so simple (see, e.g. van den Berg 2013). It turns out that experiments and practical work alone do not guarantee achieving real understanding of relevant topics. Also, as we will see later, there are other factors that can diminish the position of experiments in physics teaching and learning.

In the following text, some aspects concerning the role and possible future of experiments (especially simple experiments) will be discussed. We will not present here any review of PER studies concerning this problem though some experience mentioned below is based also on a research work done with our PhD students. Rather, some views are discussed here that can help us to look at the problem from a more general point of view. Some of them are quite obvious, while some other you can consider as rather wild. Hopefully, maybe even such "crazy ideas" can help us to see the situation from unorthodox and perhaps interesting perspectives.

As we are physicists and physics teachers, it is natural that the above-mentioned general views and ideas will be complemented by concrete examples of simple experiments usable at various levels of physics teaching and learning. In the lecture at WCPE 2, more than 30 experiments were presented; not all of them will be mentioned here not to make this text too long.

6.2 Simple Experiments: Flourishing or Endangered?

Simple experiments we use in physics teaching and learning are known under many different names: simple, low-cost, hands-on, minds-on, etc. As in many similar cases, we can remind a famous Shakespeare's quote "What's in a name? That which we call a rose by any other name would smell as sweet" (Shakespeare). It is not necessary to try to define here precisely what "simple experiments" mean; all members of the community of physics teachers and educators know from experience what this term denotes—though their particular views can differ in details, of course.

Plenty of Resources

There are plenty of resources concerning simple experiments. In days when Google is one of the main information resources for students, it is, hopefully, tolerable to use it to estimate the extent of such resources. If we try to find various combinations of the terms "simple experiments", "physics education" and "physics teaching and learning", Google informs us that there are millions of links. Asking for "hands-on experiments" returns more than 150 million results. If we put this term into quotation marks, the number of results is just 375 thousand; but even that reduced amount of web pages would be impossible to read through. "Hands-on experience in physics teaching" (not put into quotation marks) gives more than 19 million results.

We can also look at something more "solid" than web pages: at books. Amazon.com, when we ask for "low-cost-experiments", it offers more than 750 titles; if we filter the results to "physics", it is more than 70 books. Well, this seems manageable—but if we ask for "hands-on experiments" (and filter the results to "physics"), there are nearly 3000 results; "simple experiments", again filtered to "physics", provide more than 6000 books.

One could conclude from these numbers that simple experiments in physics teaching and learning are flourishing. Is it really so? A closer look can reveal some dangers.

Simple Experiments: Endangered Species?

If we put it slightly metaphorically, simple experiments may be in a position of endangered species. As in their case, the main dangers concern changing environment. By "environment", we can understand circumstances in which experiments, students and teachers coexist. Let's look at some of the aspects of such environment unfavourable to simple experiments:

Devices Around Us In the past, many devices were simple, and their function was rather clear and understandable. Consider a light bulb, an electric bell, a carbon microphone or a gas cooker. Children could easily dismount at least some of such devices, to explore how they work and, in some cases, to build at least models of such devices (e.g. it was possible to build a toy carbon microphone from three carbon sticks from old 1.5 V batteries and a paper membrane). Physical principles on which these devices were based were rather simple. Nowadays, a LED light, an electronic bell, a small condenser microphone and an induction cooker all contain electronic circuits, and their function is much more cryptic. Of course, it would be possible to "dismount" (i.e. to break) integrated circuits inside, but it would not teach us anything either about functioning of the devices or about physics.

Our Tools In the past, the tools children and pupils met (e.g. a screwdriver, a hammer, a drilling machine or a soldering iron) could do something with real objects in a physical world. Nowadays, gadgets like a keyboard, a mouse, a tablet or a smartphone are tools by which we operate with data, i.e. with objects in virtual worlds.

Simple Experiments Are No More Part of "Real Science" In the past (in the nineteenth century and before), many science experiments, even those which revealed new phenomena, were simple. Consider, for example, Faraday's experiments. Nowadays (in the twentieth and twenty-first centuries), it is not so: you cannot build LHC or LIGO in your kitchen. Therefore, simple experiments are often perceived as "just fun".

Everything Is on the Web Nowadays, all answers are, seemingly easily, "at our fingertips". It is so easy just to consume them—so why bother to make any experiments by themselves? There are many powerful rivals to simple experiments: computer games and virtual reality, information on "big science" on the Web, applets, simulations, remote labs, YouTube video clips, etc. Can simple experiments compete with them and survive?

However, one can see also a more optimistic perspective:

Simple Experiments Are Not Dead!

In the following, we will give and discuss reasons for this positive statement. We can summarize them in four points:

1. Simple does not mean stupid.
2. Simple experiments are adaptable.
3. They are "multipurpose tools" that can teach us a lot.
4. They can survive even without any purpose.

The last point can be seen as rather provocative and controversial. In fact, another "wild idea" will appear in our discussion: simple experiments spread like genes and memes. Also, in a similar parallel to evolution biology, we will coin the term "extended experiment" here when discussing advantages and disadvantages of simple experiments.

By now, it is probably clear that the author will try to support here the positive view, advocating that in spite of the above-mentioned dangers, simple experiments will survive and will play their role in future physics teaching and learning. We will return to this in the Conclusion, with both partly wild and partly cautious speculations concerning the future.

6.3 Simple Does Not Mean Stupid

By a slightly provoking term "not stupid", we mean that:

- Simple experiments can be used at various levels—from kindergarten to university. Experience shows that they are appreciated also by physics teachers at all levels. (A feedback after this lecture at WCPE 2 supported this experience, too.)
- Simple experiments need not be only quantitative. As we will see on examples below, we can use even very simple experiments for semi-quantitative and quantitative measurements.
- A lot of simple experiments can be "multilayered", which means that they can be used with various levels of theoretical explanations, details and analysis of results.

We can summarize these points by stating that simple experiments are much deeper, richer and useful for physics teaching and learning than being just means for making "wow effect" at various science shows. Let's illustrate this by a few examples.

Examples from Electrostatics: Experiments from Qualitative to Quantitative and Back

Plastic straws can be used as a superb teaching aid when we start teaching electrostatics. When charged, for example, by being rubbed with paper napkin, they attract themselves to other objects. So we can "discover" that something on the straws (that we later name charge) interacts at distance and the interaction force is attractive. Then we can continue.

Charged Plastic Straws Repel Themselves If we charge two straws and hold them in our fingers as it is shown in Fig. 6.1, we can feel that straws repel each other. (It is important to hold the straws at their ends because the straws act as levers and the force we feel by our fingers is higher than the force of electrostatic repulsion itself.) Of course, this experiment is just a qualitative one, but it can persuade us that "the same charges" repel each other, so that electrostatic interaction is not only attractive. Moreover, we also feel that the force is larger when the charges are closer to each other.

Further qualitative experiments can follow (see Dvořák 2014a), but let's turn to experiments that are not only qualitative.

How to Arrive at Coulomb's Law: A Semi-quantitative Experiment If we puncture a long plastic straw, it can turn around a pin. We charge lower part of the straw by rubbing it; the upper part serves as a pointer (Fig. 6.2).

If we bring another charged straw close to the lower part of our straw, the pointer will indicate the force between mutually repelling charges (see the left part of Fig. 6.3). We do not know the value of the force, but we can see that if we bring two charged straws (to the same place where one straw was before), the force is twice as great. We can assume that the charge of two straws is about twice as great as the

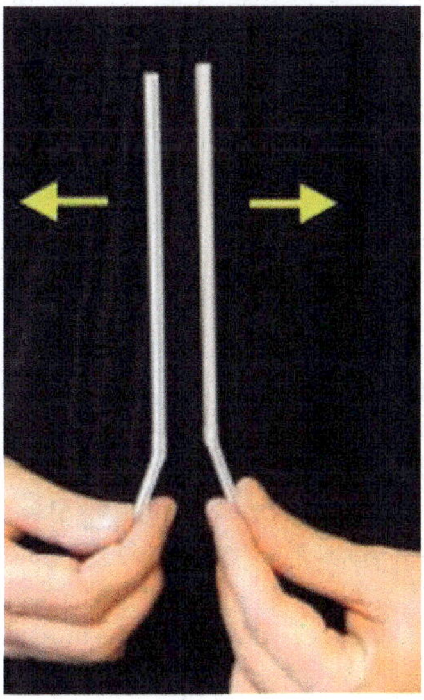

Fig. 6.1 Two charged plastic straws repel themselves and we can feel the force in our fingers

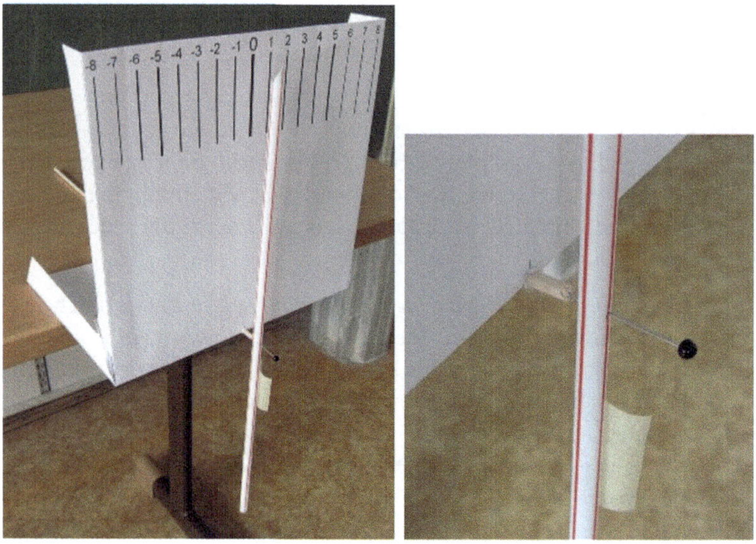

Fig. 6.2 A simple device that can measure relative values of electrostatic force

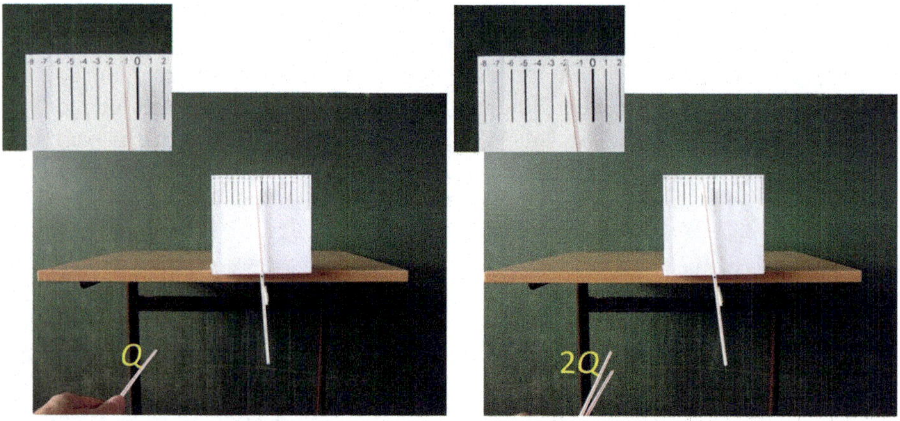

Fig. 6.3 If the charge is twice as large, the force increases two times

charge of just one straw—so we can conclude that the repelling force is proportional to the charge. (Of course, we present here the experiment and the reasoning concerning it in a very sketchy form; we do not discuss here the fact that the charges are not point-like, etc., but hopefully the basic principle is clear.)

Even more interesting is to let the charge the same and change the distance of charges. As Fig. 6.4 shows, even our simple device can be used to persuade students that halving the distance results in the force being four times as great. Therefore, instead of just stating Coulomb's law, we can "discover" it with students, at least semi-quantitatively. This is important because quite often misconceptions concerning forces between charges are quite stable in spite of formal knowledge

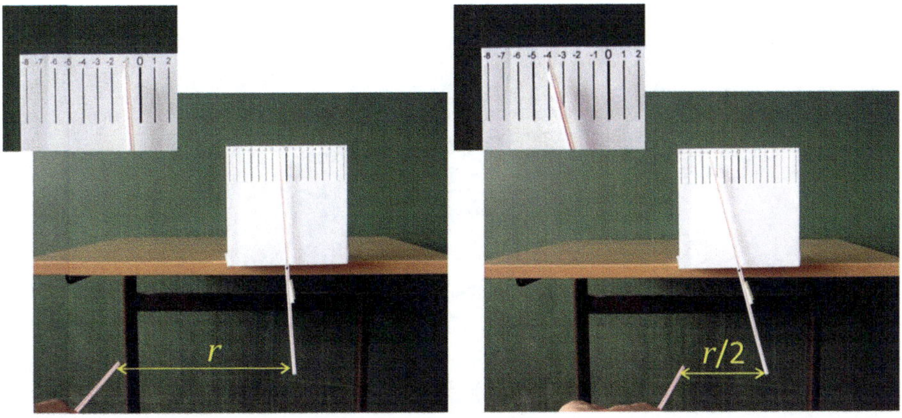

Fig. 6.4 If the distance between charges is halved, the force increases four times

of Coulomb's law. This was discussed, for example, in Koudelkova and Dvorak (2015).

Charge of a Straw: A Quantitative Experiment Previous experiments did not enable us to find the value of charges on the straws. In fact, people often have a very vague idea how large such charge can be; the estimates can vary between tens of picocoulombs and nearly milicoulombs. Laymen can tell you even more; in case you remind them that charge is measured in coulombs, they can guess that the charge of the straw is several coulombs. It can be a good starting point for a discussion about meaning of the constant k in the Coulomb's law, $F = k\,Q_1 Q_2/r^2$.

In case you hold two charged straws as it is shown in Fig. 6.5, so that the upper straw "floats" above the lower one, then the weight of the upper straw is balanced by the electrostatic repulsion. Assuming the charges of both straws are approximately the same, you can use the Coulomb's law to calculate the charges. Of course, the charges are not point-like, so the Coulomb's law gives us just a rough approximation, but the calculation can be improved (see Dvořák 2014a for details). In fact, this simple experiment can be a good motivation for students at introductory university level to apply the Gauss's law or to calculate the electric field of a finite charged line. However, perhaps a bit surprisingly, even the rough estimate using the Coulomb's law gives a value with the error less than about 50%. (Let us note that the charge of the straw is typically 20–40 nC.)

Method of Image Charges In the previous experiments, we used electrostatic repulsive force to find the charge. We can similarly use the attractive force. If we put a metal plate above a charged straw (see Fig. 6.6), it is attracted to the plate by the same force that would be exerted by another straw oppositely charged that would be placed symmetrically above the plate, as a "mirrored image" of the original straw (in case no plate is present). This experiment can be used to introduce the method of image charges for solving some problems in electrostatics.

The same effect explains quite known experiment when a light (empty) can from some drink is attracted by a charged rod. A qualitative experiment is attractive for students at secondary school level (and at undergraduate level, too). At a higher

Fig. 6.5 An experiment enabling us to find the value of charge on a straw

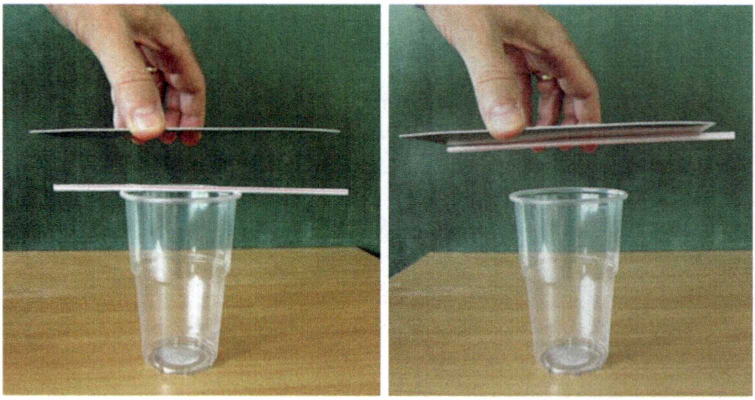

Fig. 6.6 A metal plate attracts a charged straw

level, it can be supplemented also by quantitative measurement and calculations (which, for brevity, we will not present here).

6.4 Simple Experiments Are Adaptable

"Adaptable" is a term from biology—but hopefully an analogy taken from biology can be useful when we look at simple experiments. They are "everywhere": we can find them in nearly every branch of physics (with some exceptions like string theory), at all levels of schools (from kindergarten to university), and they are attractive for all ages (and often small kids, adults and seniors are fascinated by the same experiments). Therefore, if we take it a bit metaphorically, simple experiments must be "viable".

We know from biology that animals and other species fill in all niches in a biosphere, they adapt themselves to many different environments. Similarly, we can say that simple experiments *adapt themselves* to changes of their environment. In particular, they adapt to new technologies. Let's look at some examples.

How Simple Experiments Adapt to New Technologies: From Charge to Current

In these examples we will proceed from electrostatics to electric current. Commonly available new technologies offer us a lot of interesting possibilities.

How to Measure Charge of a Straw In one of our previous experiments, we roughly established a value of charge of a plastic straw. Could it be found more precisely? Of course, it could be measured by a professional coulombmeter or by a school coulombmeters in Vernier, Pasco or similar sets—but these are "black boxes" somehow providing the value. (Try to ask your students; who of them know how these devices measure charge?) Fortunately, it is possible to measure charge also by a very simple experiment. Figure 6.7 shows both the principle and a real measurement.

The principle is simple: to measure voltage on a capacitor to which the charge is brought. (In fact, the same principle is used in coulombmeters.) $Q = C \cdot U$, so, for a capacitor with a capacity $C = 1$ μF, the voltage $U = 1$ mV corresponds to the charge 1 nC. You can see in the right part of Fig. 6.7 that our measurement gives $U = -30.8$ mV, so the charge is negative and has the value about 30 nC.

The experiment is useful in teaching and learning sequence concerning capacitors. Of course, in our simple experiment, the value of voltage drops down quite quickly as the capacitor discharges to the internal resistance of the voltmeter. (It can be later used as a motivation for behaviour of RC circuits.) Common multimeters, when measuring voltage, have internal resistance 10 MΩ, so the time constant is $RC = 10$ s, and if we read the value very quickly, in less than 1 s, the error of measured charge is less than 10%.

Charge and Capacity of a Human To discuss capacity of an isolated sphere is quite common in teaching electrostatics, but it is not very exciting. To make this topic more attractive, we can measure the capacity of a person standing on an isolating pad (e.g. a piece of Styrofoam). To do this, we connect the person with a school high voltage source to charge him or her to, say, 10 kV. Then (when disconnected from

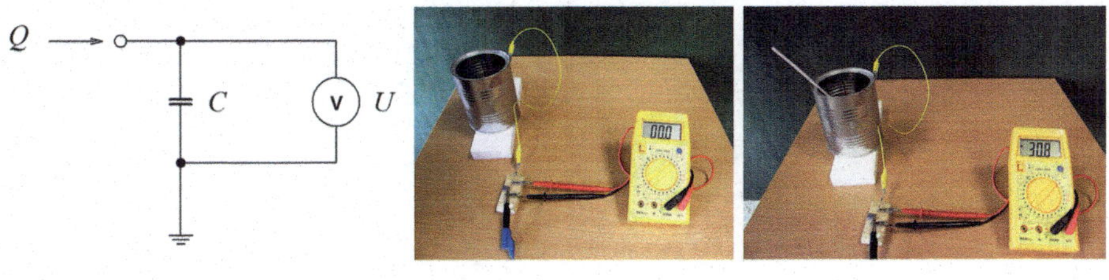

Fig. 6.7 Simple measurement of charge of a plastic straw

the source), we measure the charge of the person in the same way as we measured the charged straw. Then, using the formula $C = Q/U$, we calculate the capacity of a person. This can be done at high school level. (Students can compare who has greater capacity.) Typically, the capacity of a person is between 70 and 100 pF, for smaller kids even less. We can compare this result with the capacity of a metal sphere of radius R in a free space ($C = 4\pi\varepsilon_0 R \doteq 1.1\ \text{pF} \cdot (R/1\ \text{cm})$). We see that the capacity of the person roughly corresponds to the capacity of the sphere of a diameter equal of the height of the person.

From Electrostatics to Electric Current: Qualitative and Quantitative Experiment To demonstrate that charge can move from one conductor to another one, we can use two tins on an isolating pad connected by a piece of crepe paper as it is shown in Fig. 6.8. Charge is indicated by pieces of thin aluminium foil. (If the tin is charged, the piece of foil is repelled.) Left tin is charged; after a while we see that foils at the right tin goes up.

A piece of crepe paper serves here as a conductor—with very low conductivity—but it is just what we need to see the process of transfer of charge. Of course, many factors, for example, humidity of air, can influence this experiment. However, in good conditions it is even possible to qualitatively demonstrate how current depends on resistance and how resistance depends on parameters of a conductor: if the stripe of crepe paper is longer, the foil on the second tin rises more slowly; if we connect tins by two stripes of crepe paper, it rises faster.

Well, this experiment does not use any new technology. The following one does, and it enables to establish (or "discover") quantitative relation between charge and current—see the circuit diagram in Fig. 6.9.

If the voltage U and resistance R are high enough, the current flowing into the capacitor (that was discharged at $t = 0$) is nearly constant for some time. (9 V battery and $R = 15\ \text{k}\Omega$ suit well, the current being about 0.5 mA.) LED indicates the current, and students can see that its brightness stays more or less the same; this indicates that current does not change much. Measuring the voltage at a capacitor enables us to find the charge on it, $Q = C \cdot U$. During the experiment, we see that the voltage increases proportionally to the time t, so does the charge Q. After a short discussion, we arrive to the fact that $\Delta Q = I \cdot \Delta t$. (In our experiment, $I \doteq 0.5$ mA, and $C = 10$ mF, so the voltage rises for about 0.5 V in each 10 s; this can be easily seen and recorded by students.) Of course, as the voltage on a capacitor rises, the

Fig. 6.8 Charge transport from a left tin to a right one (using a piece of crepe paper as a conductor)

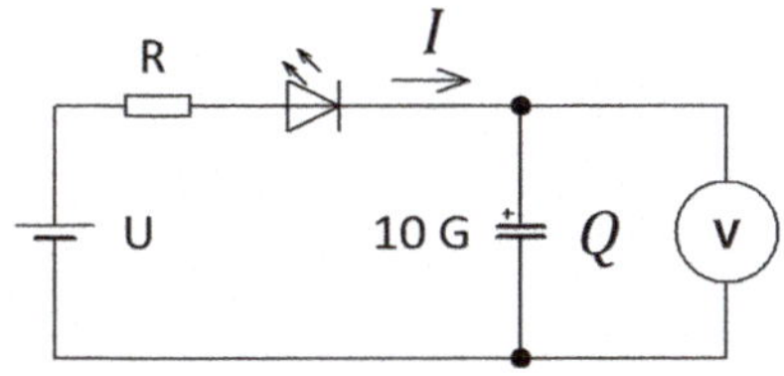

Fig. 6.9 Electric current as transport of charge

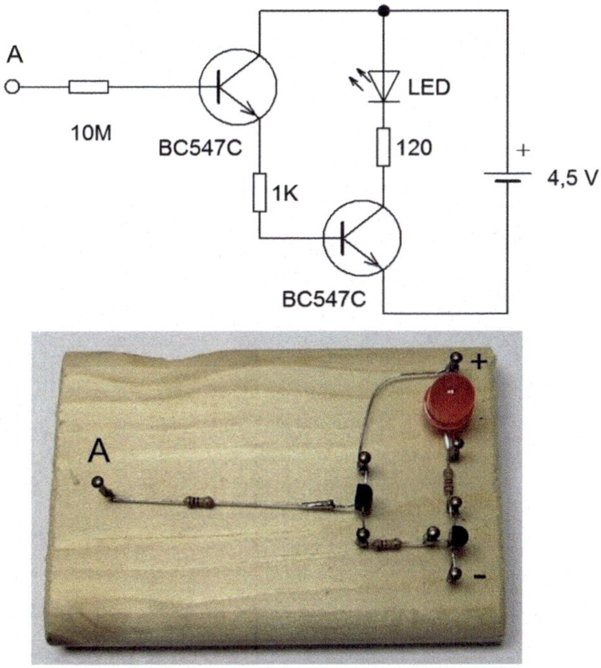

Fig. 6.10 Charge indicator with bipolar transistors

current decreases, and the relation between voltage and time is no longer linear; again, this can be discussed with students, and later it can be used as motivation for solving of the problem of charging a capacitor in an RC circuit.

Application: Charge Indicator with Bipolar Transistors A charge indicator with three bipolar transistors and a small light bulb was described in Dvořák (2012) and some experiments with it in Dvořák and Planinšič (2012). A simple version with just two transistors and LED is shown in Fig. 6.10.

The device does not indicate charge as such (i.e. it does not indicate static electric field) but changes of electric field. If we bring (negatively) charged plastic straw close to the terminal A, it attracts positive charges from outer part of wires in the device to A. (Of course, in metal wires the electrons are what is moving, but let's now speak phenomenologically about positive charges for a while.) When we move the straw apart from A, positive charges are no more attracted and flow to the base of the first transistor. This means current flowing into the base; it is amplified by the first transistor and then by the second one—and the LED shines. If we use a positively charged rod (e.g. a glass tube), the LED shines when the rod moves towards A. So, we can distinguish the polarity of charge on the rod.

This experiment proved to be very attractive both for students and teachers. In fact, many Czech physics teachers built this indicator already by themselves in seminars of the Heureka project (see Dvorakova 2014). Also, participants of the workshop at ICPE-EPEC 2013 conference built their own indicators there (see Dvořák 2014b). The device can be easily adapted, for example, to the 9 V battery (just by changing the value of the resistor in series with LED to 390 Ω), to use of PNP instead of NPN transistors, etc. In case we ground it (any terminal of the battery), connect the terminal A with a metal can standing on an insulating pad (to increase its capacity), and use a large charged plastic rod; the indicator reacts to moving the rod even at a distance of several metres.

6.5 Crazy Idea: Simple Experiments Like Genes and Memes

A while ago we used a rough biological analogy of simple experiments as "species"—maybe endangered but surely adaptable. Perhaps we can go a bit further and use even a crazier idea or metaphor, looking at simple experiments as something similar to genes or memes. Such idea can be considered as really wild, but we shall see that it can reflect some aspects of the "behaviour" of simple experiments.

Inspiration by Richard Dawkins and Umberto Eco

This metaphor is vaguely inspired by a famous book by Richard Dawkins (1976) in which he introduced the view that animals are "vehicles" enabling genes to replicate and spread. Also, he introduced the concept of memes—not physical but mental "entities"—that similarly spread among people. (If we look at Wikipedia, we can find *a meme* defined as "an idea, behaviour… that spreads from person to person within a culture". More on memes can be found in a couple of sources. One can start even from the Wikipedia page; not only supportive but also critical remarks to this concept are presented there. We do not need to dwell on details of this idea, taking it just as an inspiration.)

Similarly to what was stated above, we can look at spreading of simple experiments and formulate the following idea:

Simple experiments spread from experimentalist to experimentalist, from teacher to teacher, student, etc. within our "teaching and learning culture".

Is it too crazy? Hopefully not, at least if we do not imagine experiments as some "living entities"—which would be too much, of course. (Anyway, genes or memes are not living entities either.)

A Similar Idea: Experiments Like Books A bit surprisingly, a similar idea was stated by Umberto Eco. In one of his essays (Eco 2005), he writes that "books talk to each other". By this he means that some books influence the culture, the culture then influences the other authors, perhaps indirectly, and it is reflected by new books. In an analogy, we can say that simple experiments "talk to each other", they interact with each other, and they influence each other—of course via physicists, teachers and students who perform them.

Is This Metaphor Useful?

Surely, both analogies (genes/memes and books) are just metaphors. Nevertheless, they naturally describe, if not even explain:

- Why so many ideas of simple experiments just circulate in our community, their original authors being often unknown.
- That simple experiments spread, evolve, "mutate" and influence the culture of teaching and learning physics.
- That people who work to improve them are mostly unknown and quite often not cited (similarly to many artists and craftsmen in Middle Ages).
- That spreading and adaptation of experiments seems to be natural and inevitable process.

Let's illustrate this by a few simple experiments.

Examples: Experiments of Unknown Origin (Slightly Evolved)

Sound and Vibrations: A Simplest Experiment An experiment I used in lectures with experiments for high school students does not require any tools. I asked students how is it possible to produce some sound in case we do not have any musical instruments or tools. After a while some said that we can use our own voice. I agreed and asked them: "So, use your own voice. Please, shout!" Typically the audience is surprised by such request, but at second or third try, they shout sufficiently loud. The second step is to ask students to touch finely their neck (from the front, at a larynx) and shout again. Then I asked them to say what they felt by their hand—and many of them say "trembling" or "vibrations". So, even this childish experiment shows that sound has something common with vibrations.

Let's add just a few notes: I really know this kind of experiment from childhood, and I do not know its origin. I just added some minor points to it—introductory discussion with students, the fact that shouting and touching the necks is done in the whole group (of up to hundred students) and some elements of humour. (When asking them to touch the neck, I usually add that they can also touch the neck of their neighbour, but I warn them not to hold the other's neck too tight, and I add that it is

Fig. 6.11 How to demonstrate transversal travelling wave on a rubber rope

safer to touch just one's own neck.) Also, I put it into longer teaching-learning sequence. Surely, something similar might have been done by many other teachers and educators, so this experiment can exist in our "teaching-learning culture" in many slightly different variants.

Travelling Waves on a Rubber Rope Again, a variant of an experiment shown below was not, according to my knowledge, described elsewhere before I started to show it. However, it was surely inspired by many similar variants of experiments used to demonstrate transversal travelling waves, and perhaps it has been independently created and used by many physics teachers.

The experiment uses a piece of rubber rope (used in anoraks) about 5 m long. One of its ends is fixed (e.g. at handle of a window in class or somebody can hold it firmly). Hold the other end in one hand, push down part of the rope by the other hand and release as it is shown in Fig. 6.11.

The wave (not a harmonic one, of course, but it does not matter) travels to the fixed end, bounces, goes back, bounces at the end held by your hand, etc. Students see the wave going back and forth, and you feel the wave when it returns to your hand (it slightly pulls at your fingers). Ten returns of the wave can be perceived. It proved to be useful if the person performing the experiment counts aloud and students measure the time. (At the start, say, e.g. "Ready, steady, go!" to set the precise moment when you release the rope and students start measuring the time. When you count "Ten!", they stop their stop watches.) It is useful to have the length of the rope equal to 5 m; then the total distance travelled by the wave is 100 m. Then we can easily calculate the speed of the wave. Depending on a tension in the rubber rope, it is between about 15 and 30 m/s. You can demonstrate that for higher tension, the velocity is greater; you can also demonstrate that at fixed end, the wave is bounced with an opposite phase, etc. Moreover, the value of the speed of waves can be used also in the following experiment.

Standing Waves on a Rubber Rope We can tell our students that our rubber rope resembles a string on a guitar. However, when playing on a guitar, no one sees waves travelling back and forth. Usually, the string oscillates. This can be also modelled at our rubber rope if we move one end of it by our hand slightly up and down in the right frequency. In the lowest suitable frequency, the wavelength of standing waves is 10 m (twice the length of our rubber rope). By moving our hand twice, three times, etc. higher frequency, there are two, three or more half-waves on the rope, so the

Fig. 6.12 Standing waves on a rubber rope

wavelength λ is smaller. At a junior secondary school level, we can just qualitatively establish that higher frequency means lower wavelength. At higher level, students can measure time of ten periods of oscillations, calculate the frequency f and then discover the formula $\lambda \cdot f = \text{const.}$. (Of course, our experiment is very simple, so we should neglect some deviations.) If the tension of the rope stayed the same as in experiment with travelling waves, we can see that the constant is equal to the speed of travelling waves, so we can arrive at the formula $\lambda \cdot f = v$.

By one's hand it is possible to create no more than about 3–4 half-waves on our rope. For higher frequency, we should use some tool to generate the standing waves. A cordless drill proved to be very suitable tool. Instead of a drill bit, we put into its chunk a piece of a thick copper wire (of thickness about 3 mm) shaped to a form of a handle (see Fig. 6.12). When the drill is on, a part of the wire moves up and down and generates the oscillation of the rubber rope. (Of course, it also moves to the side, so our standing waves are not perfect, but this is sufficient for our experiments.) By changing the speed of the drill, we can generate ten or even more half-waves on our rope. Nodes and antinodes are clearly visible there; you need not define these concepts to students; they see them with their own eyes and can even touch them. A bit funny measurement can be also done using the following experiment: if we measure the wavelength (measuring distance between nodes), then, using the known velocity of waves, we can calculate the frequency—and then compare the result with what is written at the drill (e.g. 1400 rotations per minute). Similarly, we can generate standing waves on a piece of rubber rope, e.g. by an electric shaver, and measure its frequency.

Again, though a rubber rope, a cordless drill and other tools are modern ones, the origin of this experiment is quite old; probably it can be traced back to Marin Mersenne in the 1630s. Also, these experiments will surely evolve further, thanks to the creativity of many physics teachers.

6.6 Advantages and Disadvantages of Simple Experiments + Another Crazy Idea

Let's remind ourselves of some advantages and disadvantages of simple experiments in physics teaching and learning. (Our list will not be exhaustive, of course.) They are mentioned in a lot of papers and books, so we do not need to cite specific references here.

Advantages

- Simple experiments are not black boxes—it's one reason why they can help in understanding.
- They are low-cost and use simple tools.
- Some of them do not take much time.
- They work also outside school labs, in a real world.
- Students can make and do them at home.
- They are attractive and therefore can motivate students.
- They can be used for active work of students. (A famous phrase "I hear it, I forget, I see it, I remember, I do it, I understand" was already mentioned above.)

  However, it is necessary to remind these advantages are not automatic!

Weak Points

- Sometimes, simple experiments are presented as "just fun", without any discussion and explanation.
- Sometimes, the explanation is present but too shallow or even misleading.
- Sometimes they do *not* work. (Probably we all know this from our own experience; it is also summarized in a known phrase of an unknown author: "If it moves, it's biology, if it smells, it's chemistry, if it doesn't work, it's physics". In case you want a statement more specific to physics, I can offer one by the former head of our Department of Physics Education, late M. Rojko: "An experiment always works—just sometimes differently than the teacher expected".)
- Better understanding, unfortunately, is *not* guaranteed (!) This possibility was aptly expressed by R. Driver (1983) in a paraphrase of the mantra mentioned above: "I hear it, I forget, I see it, I remember, I do it, I am even more confused".

Related Crazy Idea: "Extended Experiment"

Let's take, for the last time, an inspiration from R. Dawkins. He coined the term "extended phenotype"; in fact, it is the title of one of his books. We will not discuss or analyse the meaning of this term in biology here, but it can inspire us to create a term "extended experiment". What would it mean?

A notion "extended experiment" can include, apart from the experiment itself:

- Its explanation
- Relevant theory
- Context in which we use it in teaching and learning
- Application(s) related to this or similar experiments
- Reasoning concerning the experiment
- Possible discussion in classrooms (preceding or following the experiment)

Why could such concept be useful? Because many advantages and disadvantages of simple experiments concern, in fact, *extended* experiments!

Of course, we all know that when evaluating strong and weak points of simple experiments, a broader scope is necessary (and the same is true when planning our teaching). The term "extended experiment" suggested here just explicitly names and emphasis this broader scope. Perhaps it can help us to think about simple experiments with this wider view in mind.

6.7 Experiments as Multipurpose Tools that Can Teach Us a Lot

Simple experiments can be used for various purposes, and they really can teach us and our students a lot of facts, and not only facts. Let's illustrate that again with several examples.

Experiments Can Teach Us Facts: Examples with Magnets

Magnetic Field of a Small Magnet A small magnet can be used as a good example of a magnetic dipole. Except for small distances, its magnetic field should decrease with distance as $B(r){\sim}1/r^3$. This theoretical prediction can be easily checked by a tablet or a smartphone with a magnetic field sensor (see Fig. 6.13).

We see that if we decrease the distance of a small neodymium magnet to one half, the magnetic field increases approximately eight times, in good agreement with theory. Of course, we have to measure the distance from a magnetic field sensor in our tablet or smartphone. To find this, we can use a small "trick": to magnetize a small ferromagnetic wire (an unfolded paperclip works well) and put its tip close to the tablet or smartphone where some application showing direction of the field runs. Using this, it is possible to find the position of the sensor with a precision of about 1 or 2 mm.

Force Between Magnets In introductory courses of electricity and magnetism at university level, energy density of magnetic field (in vacuum) is introduced, $w = \frac{1}{2}\vec{B}\cdot\vec{H} = B^2/(2\mu_0)$. If it is discussed only theoretically, students can often

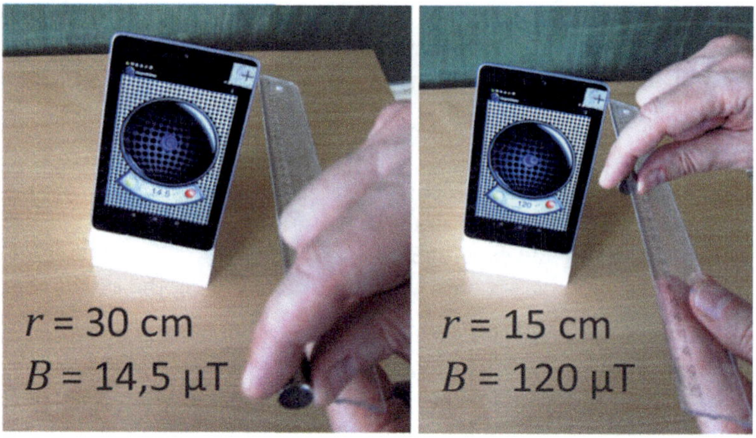

Fig. 6.13 Measuring how magnetic field of a small magnet depends on distance

perceive it as just an abstract quantity. However, we can make it very real if we use it to calculate the force by which magnets hold together in a closed magnetic circuit. As the result, $F = B^2/(2\mu_0) \cdot S$, where S is the area of magnet poles we are trying to pull apart (see, e.g. Jackson 1999, problem 5.23 at page 230). For neodymium magnets (let's estimate $B \doteq 1{,}2$ T), this gives a rather high value, more than 40 N per each square cm. This theoretical value can be verified experimentally.

A simple tool for such verification is shown in Fig. 6.14: two pieces of iron to one of which small neodymium magnets are put with their north poles alternatively pointing up and down. When we put the other iron piece to it, both stick together.

In our case the magnets had diameter 1 cm, so the total area of poles of magnets was about 12 cm^2. Therefore the total force for pulling both pieces of iron apart was more than 500 N. To demonstrate the value of this force, we can use a simple lever (a piece of a wooden board) with the experimenter himself serving as a weight (see Fig. 6.15).

The fact that results (found using one's own body) agree with a theoretical prediction can help persuading students that the energy of magnetic field is something very real.

Simple Experiments Can Teach Us Not to Be Too Proud

Simple experiments can teach us a lot more than just facts and comparison of facts and theory. I have a personal experience that is probably worth sharing as it shows how an (less carefully prepared) experiment can fail, how its results could look as a complete mystery and how a reasonable physical explanation can still be found.

It is well known that currents flowing in the same direction attract and currents in the opposite direction repel. I planned to show this in my course on electricity and magnetism in the first year of bachelor studies of future physics teachers. (This was the first year I taught that course.) I didn't want to use some professional tool where

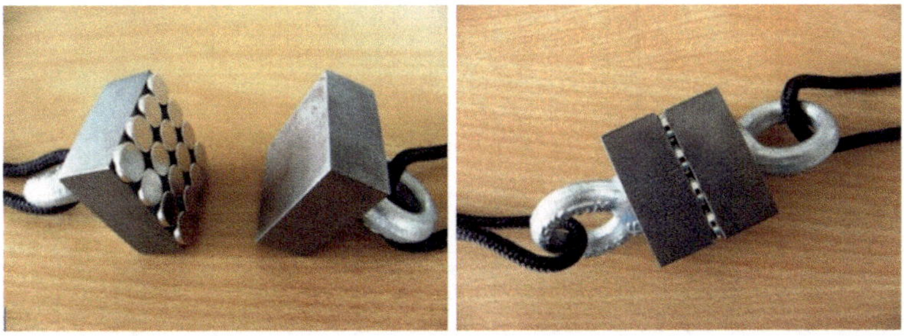

Fig. 6.14 A tool for demonstrating the force holding magnets together in a closed magnetic circuit

Fig. 6.15 A simple way how to measure (at least approximately) the force for pulling magnets apart: stepping at different points of the board changes the force

currents of order tens of amperes are present; I preferred to use simple tools students would be able to later apply in their teaching in schools. So I used thin aluminium foil stripes and a 4.5 V battery (see Fig. 6.16).

The experiment worked well for currents in the same direction; the attraction was clearly visible. However, when I showed to my students the variant with currents flowing in the opposite direction, the stripes, to my surprise, *attracted* again instead of repelling! It seemed that the experiment clearly contradicted what was a simple consequence of physics laws. (You can imagine, this is not the best moment of one's teaching career.) I had to admit I did not understand what was happening, and I would have to try to find the explanation. So, after the lecture, I repeated the experiment playing with its parameters, and finally the "mysterious" behaviour became clear. In fact, it was a slight change of configuration of the experiment that provided a clue: even when the stripes were farther apart from each other (so there mutual interaction was much weaker), they moved towards each other when the current was switched on. Such behaviour had to be caused by some external magnetic field—and it was easy to identify it as the vertical component of Earth magnetic field.

Fortunately, I was able to explain it to my students at the end of a seminar immediately following the lecture, so they could leave our department being persuaded that physics works after all. However, this experience reminded me not to be too proud when designing new variants of simple experiments and not to

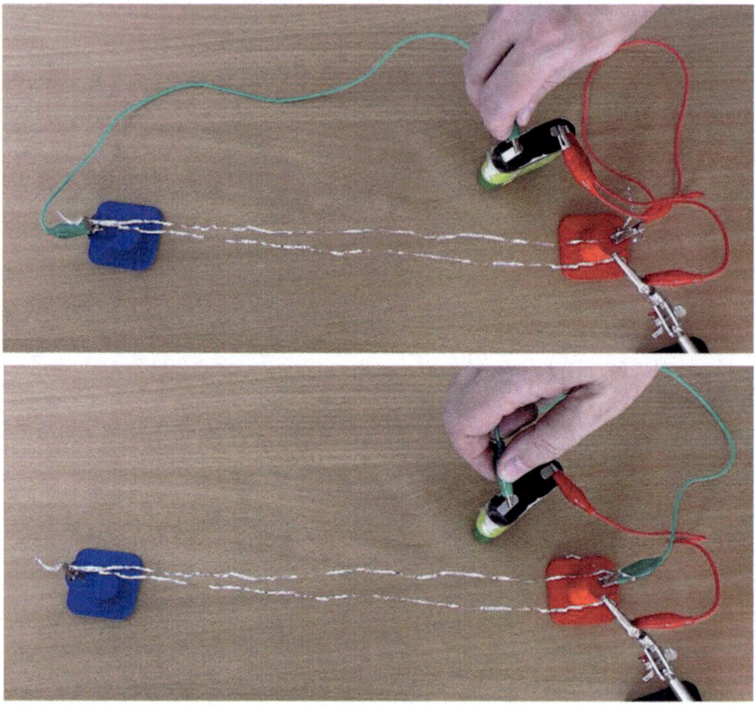

Fig. 6.16 Simple experiment that should demonstrate forces between currents can give "mysterious" results

underestimate pitfalls that can occur. (Ironically, if I used the opposite polarity of the battery, the stripes would move apart from each other, and I would be glad how well the experiment works. So, in fact, it is better that the experiment in my lecture failed so markedly and gave me a chance to learn a bit more about external influence that can spoil the expected effect.)

6.8 One More Crazy Idea: A Purpose Need Not Be Necessary

Simple Experiments Can Serve Many Purposes

Of course, many purposes of simple experiments in physics teaching and learning are mentioned in literature. Let's remind some of them (without citing concrete papers or books). Simple experiments:

- Attract attention
- Motivate

- Demonstrate physical laws
- Help to teach and learn concepts
- Develop skills
- Support inquiry based learning

So why should we speculate that they do not have to serve any purpose? Is it not a too wild and senseless idea?

Not to Serve Any Purpose: Is It Reasonable?

Well, we can turn the question in the last paragraph and ask just the opposite: Why do we always have to strive for some purpose? In fact, "purpose" refers to something other and to something more valuable. (Then, what is the purpose of this? Is there not a danger of an infinite regress here?)

However, not everything in our lives must have a purpose. In my opinion, two clear examples are music and love. They are valuable by themselves, we like them and they are part of our lives and of our culture; our life would be poorer without them. Perhaps, simple experiments can play, at least partly, a similar role. So, the last crazy idea in this paper can be specified as follows:

Simple Experiments Are Valuable by Themselves

Physics is part of our culture and experiments are part of physics. Therefore *experiments* (including simple ones) *are part of our culture*. We can enjoy them, both passively and actively, and cultivate this enjoyment, both in us and in our students. Enjoy and cultivate—it is the same as what we do in our culture with music and love.

Examples: How to Measure Gravity Acceleration

So, let's add a few experiments that could serve some purposes, of course, but even without any specified purpose, we can just enjoy them. They will concern measurement of gravity acceleration g—just sometimes in rather nontraditional ways.

By Just Dropping a Ball A free fall of a ball we drop from our lifted hand can give a very rough estimate of g. From a classical school formula $h = \frac{1}{2}gt^2$, gravitational acceleration follows as $g = 2h/t^2$. For $h = 2$ m the time of free fall is less than 1 s and more than 0.5 s. (This we can estimate just by watching a free fall, even without using a stop-watch.) So $g > 2 \cdot 2\ \mathrm{m}/(1\ \mathrm{s})^2 = 4\ \mathrm{m/s^2}$, and $g < 2 \cdot 2\ \mathrm{m}/(0,5\ \mathrm{s})^2 = 16\ \mathrm{m/s^2}$. Therefore even from this very simple experiment without practically any tools, we can conclude that g lies in the interval $\langle 4, 16 \rangle\ \mathrm{m/s^2}$. Surely, it is a very rough estimate, but it gives at least an order of magnitude of this quantity. (Well, if we take an average of the

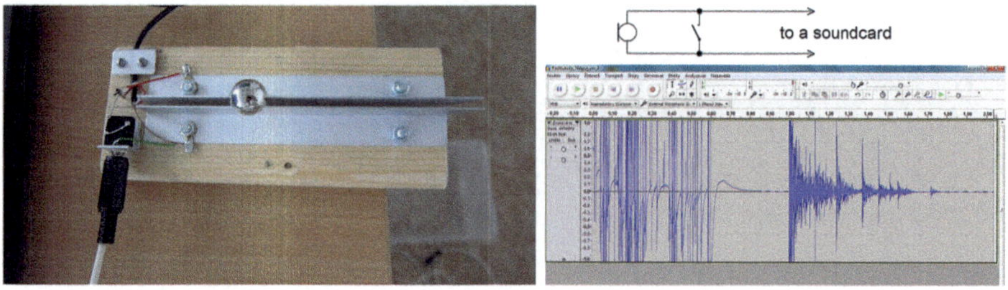

Fig. 6.17 How to measure the time of free fall of a small steel ball using a microphone attached to a computer (and software like Audacity)

values 4 and 16, we obtain 10, but we should not present this to our students as a serious way on how to determine the value of g.)

How to Measure the Time of a Free Fall More Precisely Using a Microphone and Soundcard A microphone can register the sound of fall of a small steel ball to the floor. Using a computer with a programme like Audacity, we can determine the moment of impact of the ball. To determine also the moment when the ball started to fall, we can let the ball roll on two metal rails (see Fig. 6.17).

Being on the rails, the ball short-circuits the microphone. When the ball rolls on the rails, the contact is not ideal, so there is a noise in the signal. However, the moment when the ball leaves the rails is clearly visible in the record. We do not discuss the details of the experiment here; let us just note that the setup enables to measure g even with a free fall from a height less than 10 cm!

A Nontraditional Pendulum: A Ball in a Cylinder A classical task in Theoretical Mechanics is to calculate a period of oscillation of a ball of radius r rolling back and forth in a cylinder of radius R. The result (for small amplitudes) is $T = 2\pi\sqrt{(7/5) \cdot (R - r)/g}$. Usually, the theoretical calculation is all what is done. However, we can use this result also for measuring gravitational acceleration g (see Fig. 6.18).

The sound of a ball rolling in the cylinder is recorded; the noise is largest when the speed of the ball is maximal. Software like Audacity enables us to measure the period T; the calculation of g is then straightforward.

And a Really Crazy Measurement: A Coin in a Rubber Balloon There is a well-known experiment with a coin swirling in a rubber balloon (see, e.g. The Naked Scientists 2011) or just "Google" the term "coin in a balloon experiment"; thousands of links are offered. As the coin rolls on the inner side of a balloon, the notches at the circumference of the coin bounce off of the rubber wall, and it "sings", i.e. produces a tone of a certain frequency.

Let's proceed further than just a wow effect. Knowing the number of notches of the coin, its diameter and the frequency of the sound, we can calculate the speed of the coin. Now, there is just a small step to modify this experiment for measuring g: let the coin swirl in the balloon in a circle, the axis of which is horizontal, so that the

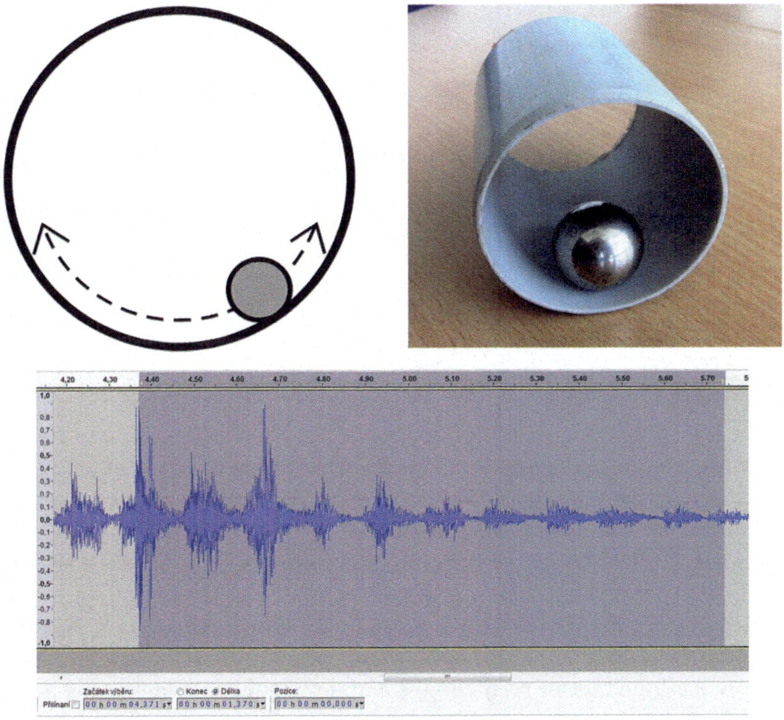

Fig. 6.18 A period of oscillations of a ball rolling in a cylinder can be measured from the sound it produces and used to determine *g*

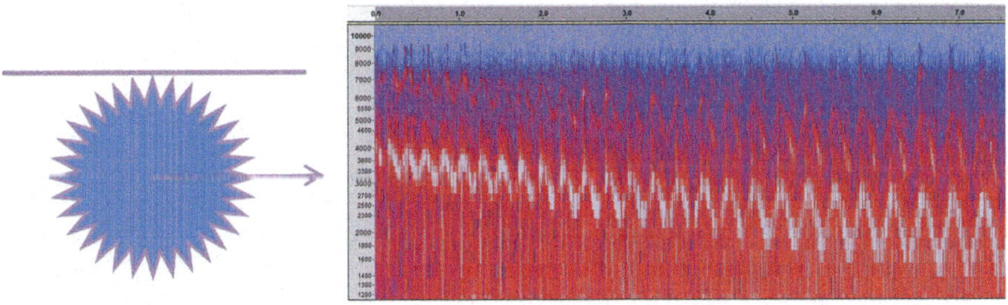

Fig. 6.19 In an experiment with a coin swirling up and down in a rubber balloon, the frequency of sound can be measured using Audacity

coin moves up and down. In the upper part, its potential energy is greater than in its lower part, the difference being $2mg(R - r)$. (The meaning of symbols is perhaps clear enough here.) This difference of potential energy is the same as the difference of kinetic energies in the lower and upper part, which can be easily calculated from the speeds of the coin—and these are determined, as was already mentioned, from frequencies. So, from the frequencies measured, e.g. again by Audacity (using a spectrum window, see Fig. 6.19), we can determine *g*.

Surely, this is a rather peculiar and funny way on how to determine gravitational acceleration, but it works (giving results about 9.6 m/s^2).

6.9 Conclusion: Perspectives of Simple Experiments

So, after presenting various views and ideas concerning simple experiments, can we dare to say something about their possible future?

It Is Difficult to Predict

First, we must remind and admit that to predict the future is a difficult and invidious task. Just consider the time scope. Our current students, future physics teachers, will teach till about 2060, more than 40 years from now. If we look back by the same time interval, i.e. to the 1970s, people were on the Moon—so cosmic expansion of man was expected. Thermonuclear power plants were predicted as a solution of energy shortage (even since the 1950s). However, there were no predictions of PCs at nearly all desks, smartphones, the Internet everywhere, etc. Just from these few examples, it is clear how hard the task of prediction is and how unreliable the results are.

In the look into our future, the uncertainties may be even greater. Will a "technological singularity" predicted by Ray Kurzweil and others arrive? Will there be a massive extended reality society envisioned, for example, in a sci-fi novel Existence by David Brin? Or what else will come? And what will be the roles of schools and teachers in different variants of future?

Opportunities for Simple Experiments

Still, if we turn back to simple experiments, we can identify some opportunities for them in the future. For example:

- New technologies will provide new tools and new materials. (We saw it in the past. For example, CDs can be used as cheap diffraction gratings. On the other hand, CDs are already declining and superseded by flash memories that are surely useless for diffraction experiments; so the change of technologies is not always beneficial for simple experiments.)
- New technologies will offer new possibilities for measurement, visualisation, etc. (We already use data loggers, IR cameras and similar instruments. What if some future glasses will enable us to see in UV region, offer the spectra, visualize measured values using extended reality, etc.? Would not it be a good tool for physics teaching and learning?)

- New technologies will bring new possibilities for spreading information. (Again, it has been here for some time already, YouTube being just one example.)

However, technology is not everything. Hopefully, some of our *human* characteristics will stay with us:

- Curiosity
- Fascination by natural phenomena
- Creativity
- Joy of mastering something by our own hands using our own minds

To Summarize

Because of new opportunities open to simple experiments and because, as we saw above, these experiments are:

- Not stupid (so they are challenging at many levels)
- Adaptable (so they spread and evolve quite naturally)
- Multipurpose tools that can teach us a lot (not only facts)
- and since we love them

we can, in my opinion, dare to predict that, similarly to music and love, *simple experiments will stay with us*—and we will enjoy them like evergreens!

References

Dawkins R. (1976): The Selfish Gene. Oxford University Press. (Note: There are many editions of this book, that from 1976 is a first one.)

Driver R. (1983): The Pupil as Scientist? Open Univ. Press, Milton Keynes. (Cited, for example, in Monk M., Osborne J. (2000): Good practice in science teaching. What research has to say. Open Univ. Press, Maidenhead.)

Dvořák L. (2012): Bipolar transistors can detect charge in electrostatic experiments. Phys. Educ. 47 (No. 4), 434–438

Dvořák, L. (2014a): Simple quantitative electrostatic experiments for teachers and students. In: Proceedings of selected papers of the GIREP-ICPE-MPTL International Conference Reims, August 22–27, 2010. Eds: W. Kaminski, M. Michelini. Università degli Studi di Udine. LithoStampa – Pasian di Prato (Udine) 2014. ISBN 978-88-97311-32-4, pp. 355–361.

Dvořák L. (2014b): Semiconductors at Work. In: ICPE-EPEC 2013 Conference Proceedings. Editors: Dvořák L and Koudelková V. Charles University in Prague, MATFYZPRESS publisher, Prague, 2014. ISBN 978-80-7378-266-5. pp. 818–825. Available online at http://www.icpe2013.org/uploads/ICPE-EPEC_2013_ConferenceProceedings.pdf

Dvořák L, Planinšič G. (2012): Experiments with charge indicator based on bipolar transistors. Phys. Educ. 47, (No. 6), pp. 721–727

Dvořáková, I. (2014): Active learning in the Heureka Project – teachers in the role of students. In: ICPE-EPEC 2013 Conference Proceedings. Editors: Dvořák L and Koudelková V. Charles University in Prague, MATFYZPRESS publisher, Prague, 2014. ISBN 978-80-7378-266-5.

pp. 47–62. Available online at http://www.icpe2013.org/uploads/ICPE-EPEC_2013_ConferenceProceedings.pdf
Eco U. (2005): On Literature. Harvest Books. ISBN 0156032392. (There are various editions of English translation of this book, even Kindle edition. The idea of books "talking" to each other is mentioned in the chapter Borges and My Anxiety of Influence.)
Jackson J.D. (1999): Classical Electrodynamics. Third Edition, John Willey& Sons.
Koudelkova, V., Dvorak, L. (2015): High school students' misconceptions in electricity and magnetism and some experiments that can help students to reduce them. Il Nuovo Cimento C (3). DOI: https://doi.org/10.1393/ncc/i2015-15101-7
Shakespeare W.: (2018). Romeo and Juliet. Available online at MIT's website: http://shakespeare.mit.edu/romeo_juliet/romeo_juliet.2.2.html.
The Naked Scientists (2011): Roaring Baloon. Available online: https://www.thenakedscientists.com/get-naked/experiments/roaring-balloon
van den Berg E. (2013): The PCK of Laboratory Teaching: Turning Manipulation of Equipment into Manipulation of Ideas. Scientia in Educatione 4 (No. 2), 74–92. Available online: http://www.scied.cz/index.php/scied/article/viewFile/86/72

Part III
Research-Based Alternatives to Traditional Physics

Chapter 7
Research-Based Alternatives to Traditional Physics Teaching at University and College

Jenaro Guisasola

Abstract The instructional alternative approaches to the traditional lectures in physics education are a growing area of research and development for improving physics teaching. Physics education research (PER) proposes new teaching approaches in a gradual research-based evolutionary process aiming empirical development at iterative development. These approaches may study didactical transformations of scientific content, design of well-documented sequences of teaching activities and students' learning progression monitored by various methods, produce evidence-based innovative products and so on. In this chapter, I introduce the context of the organization of the Round Table of GIREP Thematic Group Physics Education Research at University (GTG-PERU) developed in the 2nd Word Conference on Physics Education. I provide an overview of trends with regard to different interactive methodologies of instruction and analysis of students' learning.

University education models for science and technology are being analysed and questioned throughout the world. Although the majority of scientists, mathematicians and engineers successfully managed to learn within a traditional teaching format, they are the exception and not the rule. University-level scientific-technological education should support a diverse student population where actually using knowledge, not just memorizing it, is becoming more important. Bearing in mind that Europe has some of the best universities in the world, the report to the European Commission on improving the quality of teaching and learning in Europe's higher education institutions (2013) advocates introducing a new university education focus that differs radically from traditional pedagogic methods. The report states that "the public authorities responsible for Higher Education should ensure the existence of a sustainable, well-funded framework to support higher education institutions' efforts to improve the quality of teaching and learning". The group of experts that wrote the report are asking political leaders from all over Europe to implement the change by adopting scientific education focussed on investigation and problem-based learning

J. Guisasola (✉)
University of the Basque Country (UPV/EHU), San Sebastian, Spain
e-mail: jenaro.guisasola@ehu.es

M. Pietrocola (ed.), *Upgrading Physics Education to Meet the Needs of Society*,
https://doi.org/10.1007/978-3-319-96163-7_7

(PBL). Similar calls are being made in other countries such as the USA where the *Guide to Implementing the Next Generation Science Standards* project defines and justifies aims and procedures for scientific education, promoted by the National Research Council (2015).

Compared with traditional courses that transmit knowledge, active teaching can improve the level of learning concepts and laws and, of course, the skills required to apply the knowledge in different contexts. Teaching based on *active methodologies* (PBL, projects, learning by inquiry, etc.) helps students to use deep, transferable understanding, way beyond the end of the course. As a consequence of the changes on teaching strategies, in recent years, changes have been made in curriculum content, which has been to carefully select the central ideas and concepts running through the physics degree syllabus. Scientists and engineers possess a great quantity of knowledge organized around key concepts. Students should develop knowledge in relation to key ideas in the disciplines, instead of learning many different ideas and techniques. The ideas should be developed during the degree course by means of carefully designed learning activities, and the assessments should provide students with feedback and learning. Regular university courses usually provide ideas and practices compartmentalized into chapters that blur connections within the discipline and with other disciplines in the degree course and make it difficult for students to correlate facts, ideas and scientific practices. On the other hand, focussing the basic ideas from each discipline helps to reduce the quantity of material that, as many agree, has become overwhelming. This lack of selection frequently leads to "mile-long" syllabuses offering merely superficial reflection ("a mile wide but only an inch deep" as the Spanish saying goes).

In the changes and initiatives described, science teaching research has developed a key role, particularly research into physics teaching (PER). Over the last few decades, PER has detected student difficulties in terms of understanding concepts and applying knowledge in different situations. Studies have revealed a wide gap between what teachers wish to teach and what students really learn (McDermott 2001). However, PER goes beyond identifying deficiencies in student learning and traditional teaching. Researchers have worked hard to present solid teaching proposals (Hake 1998). PER compiles the enormous effort put in by PER researchers and physics teachers to present instruction materials and methods that have been fundamental in its design and repeatedly assessed. However, deficiencies persist even in the most recent educational focus points, and many questions remain unanswered. In this chapter, we will present well-founded studies that endorse changes recommended by physics teaching research and present educational innovations.

The instructional alternative approaches to the traditional lectures in physics education are a flourishing area of research and development for improving physics teaching. PER proposes new teaching approaches in a gradual research-based evolutionary process aiming empirical development at iterative development. These approaches may study didactical transformations of scientific content, design of well-documented sequences of teaching activities and students' learning progression monitored by various methods, produce evidence-based innovative products and so on.

In the next chapters, from the Round Table of GIREP Thematic Group Physics Education Research at University (GTG-PERU) developed in the second World Conference on Physics Education, we provide an overview of trends with regard to different interactive methodologies of instruction and analysis of students' learning. Mila Kryjevskaia's study is a project focused on analysing the students' reasoning after the instruction. The study analyses, from the psychological theory called "dual process", student reasoning in physics. In relation to the results of the study, it is worth noting the confirmation of the persistence and relevance of the incorrect reasoning, intuitively often employed by introductory physics students. It should be highlighted that these forms of incorrect reasoning may not necessarily be based on everyday ideas related to a situation at hand. Many students found confirmation of their intuitive ideas in misinterpreted formal mathematical relationships.

The studies presented by Ruben Limiñana et al., Marisa Michelini and Alberto Stefanel and Genaro Zabala discuss some teaching strategies for improving students understanding of physics in different university degrees such as primary teacher degrees, bio-area degrees and engineering degrees.

Ruben Limiñana and colleagues present a study focused on teaching strategies based on an inquiry-based teaching model, but it is about fundamental problems of science (oriented research about fundamental problems of science) in this case, about solving the question on where are we on a spherical Earth. The definitions of latitude and longitude appear from solving the location of human beings on the Earth. It is necessary to point out that the studio shows the activities that are done in a classroom and that can be interesting for the readers. Also, it indicates that the students improve significantly in the learning of the concepts of latitude and longitude, as well as the abilities to apply them to the knowledge of where are we on a spherical Earth.

Marisa Michelini and Alberto Stefanel present a project on research-based intervention modules that are developed in the bio-area degrees in the University of Udine. For designing the intervention modules, the study, firstly, takes into account how to integrate physics into biology curriculum (or biology into physics curriculum) beyond simple provision of examples from the respective disciplines. That is, the project redesigns the way in which physics is offered so that its role can be recognized in the specific subject matter characterizing the degree: turning the ways in which physics is approached, changing the role of each topical areas, and individuating specific applications of physics in the professional field of the degree. Secondly, the teaching proposal addressed the role for an integrate learning of methodological aspects, such as problem solving and modelling. The chapter shows the designing, activities and evaluation of the module on fluid. The different activities and evaluation questions that are showed will be interesting for the readers. Finally, the results obtained on students' learning outcomes indicate the effectiveness of the proposal.

Genaro Zavala's study works on the design of problems based on cognitive scaffolding to teach physics. These problems are designed to be used in almost any setting since no equipment is needed. Students work in collaborative groups of three or four students each. The design consists on transforming a traditional problem to a tutorial-format problem which takes the student through scientific reasoning steps to build concepts, that is, cognitive scaffolding. The study provides some examples and results on improving students understanding.

The discussion about research-based science teaching made by Cristiano Mattos thinks over a general consideration on teaching science and concrete condition to implement research-based teaching. The author discusses, in the context of university physics teaching, about teacher styles, nature of physics, nature of teaching process and values and ends of scientific knowledge. It discusses the need or not of a global curriculum and curriculum evaluation.

To finish off this introductory chapter to the studies from Round Table of GIREP Thematic Group Physics Education Research at University (GTG-PERU), I would like to highlight the importance of the research into learning and teaching physics presented in the chapter, as developed from the physics departments. We are not suggesting that PER should not be carried out in the schools of education, but we do not believe that the PER area is feasible without a critical mass of teachers in physics departments. Science education research has mainly been developed for primary and secondary levels and has had a low impact on post-compulsory secondary levels (16–18 years old) and on university courses. This can be explained by stating that research into education performed by physicists from the physics department is more accessible and, in general, more relevant for the physics teaching staff than research performed in the schools of education. PER is a single enterprise where the research methods are strongly influenced by physics discipline and that uses results finds of other disciplines that deal with the cognition and learning (cognitive sciences). A "discipline-based educational research (DBR)" (2012) movement has emerged recently advocating educational research projects within the departments of the scientific disciplines with the main aim of researching and developing teaching methods and evaluating their effectiveness so that teaching staff in the scientific disciplines might use them. DBER scholars have devoted considerable attention to effective instructional strategies and to students' conceptual understanding, problem solving and use of representations.

In the next chapters, we provide an overview of trends with regard to different interactive methodologies of instruction and analysis of students' learning. We discuss teaching approach frameworks and their features across different characteristics, such as transformation of the content, explicit monitoring of students' learning and evaluation.

References

Hake, R.R. (1998) Interactive-engagement versus traditional methods: A six-thousand-student survey of mechanics test data for introductory physics courses, American Journal of Physics. 66, 64–74.

McDermott, L.C. (2001). Oersted Medal lecture 2001: Physics Education Research-The key to student learning, American Journal of Physics. 69, 1127–1137.

National Research Council. (2015). *Guide to Implementing the Next Generation Science Standards*. Committee on Guidance on Implementing the Next Generation Science Standards. Board on

Science Education, Division of Behavioral and Social Sciences and Education, Washington, DC: The National Academies Press.
Report to the European Commission on Improving the quality of teaching and learning in Europe's higher education institutions. (2013). ISBN 978-92-79-30360-9. In: http://ec.europa.eu/dgs/education_culture/repository/education/library/reports/modernisation_en.pd

Chapter 8
A Reflection on Research-Based Alternatives of Physics Teaching on Educational Activity System

Cristiano Mattos

Abstract In this chapter, we present a general reflection on the place of research-based alternatives compared to traditional physics teaching to professional formation, seeking to establish a broader dialogue with the works presented in the previous texts of this part of the book. Our main purpose is to localize alternative teaching proposals in the education's activity chain, since higher education activity is part of a larger educational system that connects basic school to productive working life. We indeed bring more questions than real answers since school-society relation has been the object of research for a long time in several research fields. We assume a radical position, seeking to grasp the root of this matter, thinking critically about the concrete conditions of the research-based teaching approaches, highlighting some points and presenting a preliminary overview of what could be a common ground to think about the research in physical teaching in higher education.

8.1 Introduction

We all know that any well-based knowledge is crucial for solving real-life problems and then well-based researches in teaching methods are welcome to any level of education. Its validation reflects the sincere efforts of researchers to produce better teaching and learning. However, the core issue of all educational activity is clarifying its own objective, i.e. what do we intend to teach? How do we intend to teach? And why do we intend to teach? The objective reflects the complexity of the activity that includes the coordination of several minor objectives constituting a hierarchically complex of objectives. Therefore, local actions, e.g. the application of new teaching methods, have local objectives such as learning a concept, a procedure or an argument. However, a discipline is just one of the coordinated actions of a larger activity (e.g. undergraduate course) that have specific purposes that must be specifically and differently coordinated to educate a physicist, an engineer or a physician.

C. Mattos (✉)
Institute of Physics, University of São Paulo, São Paulo, Brazil
e-mail: mattos@if.usp.br

M. Pietrocola (ed.), *Upgrading Physics Education to Meet the Needs of Society*,
https://doi.org/10.1007/978-3-319-96163-7_8

The same rationale can be applied when we understand the need of coordination between the university and the institutions of work placement. The objectives of both institutions must be coordinated in order to be considered as a professional training unit. Here we identify the curriculum as a possible unit of analysis of education that allows identifying the internal coordination levels of an entire course, even those objectives society give financial support.

In this way, the meaning of research-based alternatives to traditional physics teaching emerges in the relations established within a complex educational activity chain that involves persons. In addition to the institutional coordination established, the meaning of a new method of teaching will depend on the theoretical and methodological foundations, the conceptions of human being, human relationship and scientific knowledge it is based on. All aspects related to the epistemological, ontological and axiological dimensions reflect the choices made by the proposers of the new method.

8.2 Higher Education Teaching Cha(lle)nge

Researches to support teaching methods are fundamental, mainly because considering education as human science could be seen in a pre-paradigmatic period, since there are a dozen of theoretical proposals to education or science education.

Nevertheless, a common ground on different physics education theoretical perspectives is that to teach physics, in any level, but particularly in higher education, a teacher must have a deep knowledge of physics. However, at the same time, we already know that this scientific proficiency is not enough to ensure students' learning or engagement. Besides mastering on scientific subjects, science or engineering or medical people should understand what meaning the scientific knowledge they master would have in the social environment they will work in.

At the same time, independently of which profession a student pursue, he will have to establish educational relationships that are intertwined in an educational process of convincing others about the validity of the knowledge he wants to use as a solution to the problems for which their expertise has been called to account. Then, in diverse life situations, such professionals could become teachers with different styles, such as authoritarian, permissive or authoritative, determining the types of relationships with their future colleagues or students (Chamundeswari 2013), which may facilitate their engagements in the productive processes for which they are called to participate in, whether at a school, university or industry.

8.3 Concrete Conditions to Sustain an Educational Change

Traditional physics teaching is usually based on the assumption that the successful teaching is the one where the physics content is reproduced specifically on problem-solving activities (Ceberio et al. 2008). Moreover, in higher education, particularly in physics courses, many ontological, epistemological and axiological assumptions about physics are crystalized. Very often teacher and students at physics courses, for instance, do not use to debate the nature of science or the role of history of science (Gooday et al. 2008; Höttecke and Silva 2011) to understand the pragmatic achievements of science. This approach usually emerges in cultures where the scientific knowledge is considered True (with capital T).

This kind of epistemological and ontological commitments reflects students' and teachers' alienation from a broader complex relation the scientific knowledge has with social and cultural human life. In general, this view of science leads teaching to a banking education or a teacher-centred education method, where the value and ends of knowledge to be taught and learned seem to be not just predefined but also crystalized.

Therefore, in order to implement higher education alternative teaching methods, the proposers must face not only the resistance of students and colleagues but also the resistance of institutionalized traditional teaching cultures (e.g. Hernandez et al. 2014). Such institutionalization could be found in different levels of the educational activity, educational boards, committees and "unifying educational programs", which reproduce the epistemological idea that physics science is one and well defined, the ontological idea that the nature exists independently of the humankind and the axiological idea that the value of scientific knowledge is major for human well-being. Unfortunately, there is no room here for a deeper discussion regarding these assumptions. In spite of these ideas being widespread, we barely have an institutional space to discuss such commitments that lead to "ways-of-being scientific".

One of the consequences of those commitments is that facing institutional and personal resistance to understand that science education belongs to the field of human sciences; many teaching experiments were lost seeking for "hard science" recognition, trying to be validated as a physical experiment (Schultz 2001; Handelsman et al. 2004). Beyond that, several institutions have highly bureaucratic steps to implement new syllabus or curricula allowing pavement to new teaching methods.

8.4 Generality of the Research

We assume that new teaching methods would be an answer for educational demands. However, one of the biggest problems to generalize human science methods, particularly in education, is to know if the demands are the same in different communities or countries, with different culture and social needs. This assumption put in check the idea that the content or the teaching method should be the same in all contexts.

Considering specific teaching methods as special cases of a more general method, we have to establish what "general" means. Is it a method applicable to any sociocultural context, regardless of any concrete conditions for implementation? Is it a students' and teachers' proof method? Or should they be methods with social contexts domains of validity, where each domain is different considering different commitments (Burchianti and Barrero 2016)? Such commitments or principles refer to the nature of the scientific knowledge, the "methods" used to produce the knowledge and the values and ends we attribute to this knowledge. In an interdisciplinary research area such as science education, one of our biggest challenges is to clarify what commitments we are assuming, allowing others to understand what are the principles our educational innovations are based.

The previous texts in this chapter presented research-based alternatives to traditional physics teaching. Their objects are the physics teaching in higher education for engineering courses, teacher formation, bioscience areas and teaching to novice and almost undergraduate students. They are using different concepts, such as metacognition, intuition, cognitive scaffolding and different interactive methods such as student working groups, individual strategies, etc.

All researches are well based and have determined the context of validity of its results. However, for instance, no one has asked what is the role of students' professional choice in learning or what is the meaning or the value students give to science. Those points are usually default and well stated. It seems reasonable to think that these points can determine different engagements with knowledge in the discipline, in the whole course, or can determine the will for a productive engagement in all education activities.

For example, generally we could agree that teaching children in an oncological hospital class, or teaching people without material support, or teaching those who don't want to learn, or even teaching in war situation, demands quite different ways to teach (Mattos and Tavares 2014; Wattar 2014; Nathan 2006; Johnson 1944). Far beyond, all those situations demand clearly understanding of what purpose these persons have with the knowledge they are willing to learn, meaning that teachers should have an ethical agency to grasp the axiological dimension of learners' intentions.

According to our observations about Brazilian preservice physics teacher's profile, we know that most of them accept to be a teacher only late in their training course. On the other hand, engineering students often realize, after their undergraduation, that the profession they were in is not the one they would like to work (Fiske 1996). However, most of our teaching methods take as granted that students' career choice is definitive in the beginning of their university studies.

Of course, it is possible to identify many students productively engaged, but most of them have illusions about their careers and about how they will experience the use of the specific contents throughout their lives (Wyer 2003; Tan-Wilson and Stamp 2015). Nevertheless, alternative student-centred methods are needed to take place of those teacher-centred that are fading away students' illusions about their careers, giving hopeless perspective to profession as a joyful life.

8.5 Curriculum

In the last years, for different reasons, efforts to unify curricula in Europe, the United States and Brazil have been made, most of them trying to build a more global curriculum from basic school to the university levels. However, we are facing contradictions based on the dualism between local and global forces disputing space on curricular objectives and contents (Vulliamy 2010; Wattar 2014).

Curricular transformation is a necessity when we understand that societies are transforming and new demands emerge from different economic and social fields (Carnoy 1999; Haste 2010). At the same time, in many physics education institutions, curricula are untouched. One of the main evidences of contents' immobility in physics teaching practice is the stability of textbook contents. There are assumptions about the nature of taught knowledge unchanged in the best sellers' textbooks used all over the world, such as "Halliday", "Tippler", "Goldstein" and "Callen" among others. These textbooks present the same set of specific content with minor differences, structuring a worldwide standard physics list of contents (use to be called as curriculum). The problem is that the teaching method subsumed in these textbooks usually reinforces traditional teachers' understanding that there is only one way to teach the contents, which is the way they had learned physics.

Fortunately, contrary to these beliefs, the researches previously presented on this chapter introduced a diversity of methods. Michellini and Stefanel pointed out the central ideas that Peru Group is trying to achieve, interpreted by Guisasola (cross-reference) as the need for:

> turning the ways in which physics is approached, changing the role of each topical areas, individuating specific application of physics in the professional filed of the degree, and offer instruments and methods building the different fields.

The excerpt indicates the assumption that physics must be taught not only by different methods but also for different physics training needs or different educational objectives.

Despite the local advances made in the previous work of my colleagues, we can say that we are in a pre-paradigmatic moment where innovations dispute a place in the sun, trying to be viewed as more successful than others are. The result in the last years is the proliferation of a soup of letters that are popping up all over the world as acronyms, for instance, SCALE-UP (Student-Centered Activity Learning Environment with Upside-Down Pedagogies—Beichner et al. 2007), PI (Peer Instruction—Crouch et al., 2007), PUM (Physics Union Mathematics), REACT (Relating, Experiencing, Applying, Cooperating, Transferring—Crawford 2001), JiTT (Just-in-Time Teaching—Novak et al. 1999) and Isle (Investigative Science Learning Environment—Etkina and Van Heuvelen 2007), among others.

8.6 Evaluation

Thus, it is imperative to advance in developing evaluation criteria to create a common ground criteria should be created, not as a way to force a unified teaching method or curriculum such those the multinational educational corporations would like to establish all over the world, treating educational problems with the same logic of economic problems.

These assessments should be carried out throughout various hierarchical levels of the education system from the immediate learning concept with local problems presented in classroom situation to long-term evaluations throughout all undergraduation courses and beyond school walls reaching in the wild.

It is important that the evaluation overcomes the school walls reaching people's life, but it is paramount to evaluate our capacity to live together, as a social being, dimensioning the role of the scientific knowledge to construct a compassionate life to the humankind. It is essential to overcome dehumanization structures of power that treat each individual as an expendable unity of a depersonalized whole.

Considering the social system as a complex one, each hierarchical level brings different epistemological, ontological and axiological commitments. To overcome students' simple engagement, a productive engagement should be aimed at different hierarchical levels, in a way that students do not only consume knowledge but also produce knowledge collectively. This tension between consumption and production of knowledge leads us to question what problems are we offering to our students? Are these problems engageable? Schools' mission should be to introduce and highlight the social contradictions and to overcome dichotomies such as school's indoor versus outdoor problems, individual versus collective problems, personal versus political problems and local versus global problems.

In order to overcome these dichotomies, students should create meaningful relationships between the immediate specific content and curriculum objectives, allowing then to connect an immediate learned concept with the conceptual ecology required to understand its role in the construction of alternative social concrete conditions for a democratic and respectful society.

8.7 Conclusion

Coherently with our initial proposition, we are bringing more questions than solutions. We do not work with the idea of a general method. From the perspective of dialectical materialism, a method is an "epistemology within an ontology"; in other words, the objects emerge within the means we know it (Rodrigues et al. 2014). Then, beyond the generality of science, science educational methods should take into account local demands and its relation with global ones. Furthermore, the science education methods and their associated evaluation criteria should consider learning as a lifelong learning (Longworth 2003).

The problem of how could people educate others through all their lives had driven us to consider education in a broader sense, where a critical learning society accomplishes education in all hierarchical levels of the social system (Welton 2005; Arlow 1999; Sundström and Fernández 2013; Wells 2010). Then, facing social problems, teaching responsibilities should came not just with the individual teachers' practice but also with an institutional structure to support the needed institutional changes to implement sustainable research-based teaching activities.

The reflection we made led us to more questions: when we think education as a complex hierarchical system encompassing classrooms, schools, societies, nations and cultures, what should be "conserved and transformed" in physics teaching activities? Should teachers and engenieers learn the same physics using the same methods? What are the collective learning difficulties? What is the nature of the learning difficulty: cognitive, social, historical or cultural? Can the difficulty be attributed to the teacher-students relationship, such as different power relations? When are teacher and student co-responsible for the teaching-learning process? What are the responsibilities of the university in the civilizational process? (Arthur and Bohlin 2005).

Mila Kryjevskaia (cross-reference) wrote, "it is a common expectation that, after instruction, students will consciously and systematically construct chains of reasoning that start from established scientific principles and lead to well-justified problems". Escalating the expectations, we hope that these chains of reasoning should go further in the higher levels of the hierarchical chain of problems that connect students' immediate life to broader social-historical-political problems, reinforcing commitments and complexifying the consciousness of their role in the humanization process.

References

Arlow, M. (1999) Citizenship education in a contested society, *The Development Education Journal*, vol. 6.1, October.

Arthur, J.; Bohlin, K. (eds.) (2005) *Citizenship and higher education: the role of universities in communities and society*. London: RoutledgeFalmer, 2005.

Beichner, R. J.; Saul, J. M.; Abbott, D. S.; Morse, J. J.; Deardorff, D. L.; Allain, R. J.; Bonham, S. W.; Dancy, M. H.; Risley, J. S. 2007. The student-centered activities for large enrollment undergraduate programs (SCALE-UP) project. In *Research-Based Reform of University Physics*, edited by E. F. Redish and P. J. Cooney (American Association of Physics Teachers, College Park, MD, 2007), Reviews in PER Vol. 1, Available at: http://www.percentral.org/document/ServeFile.cfm?ID=4517

Burchianti, F.; Barrero, R.Z. (2016) The Controversy about Education for Citizenship: The Contested Limits of Tolerance in Spain. *Societàmutamentopolitica*, 7(13), pp. 269–287.

Carnoy, M. (1999). *Globalization and educational reform: what planners need to know*. Paris: UNESCO, International Institute for Educational Planning. Retrieved from http://unesco.amu.edu.pl/pdf/Carnoy.pdf

Ceberio, M.; Guisasola, J.; Almudí, J.M. (2008) ¿Cúales son las innovaciones didácticas que propone la investigación en resolución de problemas de física y qué resultados alcanzan? *Enseñanza de las ciencias*, 26(3), pp. 419–430.

Chamundeswari, S. (2013) Teacher Management Styles and their Influence on Performance and Leadership Development among Students at the Secondary Level. *International Journal of Academic Research in Progressive Education and Development*, 2(1), pp. 367–418.

Crawford, L. M. (2001) *Teaching Contextually*: Research, Rationale, and Techniques for Improving Student Motivation and Achievement. Texas: CCI Publishing, Inc.

Crouch, C. H., Watkins, J., Fagen, A.P., & Mazur, E. (2007). Peer instruction: Engaging students one-on-one, all at once, in *Research-Based Reform of University Physics*, edited by E. F. Redish and P. J. Cooney (American Association of Physics Teachers, College Park, MD, 2007), Reviews in PER Vol. 1, Available at: http://www.percentral.org/document/ServeFile.cfm?ID=4990

Etkina, E., & Van Heuvelen, A. (2007). Investigative Science Learning Environment – A Science Process Approach to Learning Physics. in *Research-Based Reform of University Physics*, edited by E. F. Redish and P. J. Cooney (American Association of Physics Teachers, College Park, MD, 2007), Reviews in PER Vol. 1, Available at: http://www.compadre.org/Repository/document/ServeFile.cfm?ID=4988&DocID=239

Fiske, P. (1996) An Open Letter To Frustrated Scientists Looking For A Job: There Is Hope. The Scientist, 10(20). Available at: http://www.the-scientist.com/?articles.view/articleNo/18101/

Gooday, G.; Lynch, J.M.; Wilson, K.G.; Barsky, C.K. (2008) Does Science Education Need the History of Science? *Isis*, 99(2), pp. 322–330.

Handelsman, J.; May, D.E.; Beichner, R.; Bruns, P.; Chang, A.; DeHaan, R.; Gentile, J.; Lauffer, S.; Stewart, J.; Tilghman, S.M.; Wood, W.B. (2004) Scientific Teaching. *Science* 23, 304(5670), pp. 521–522.

Haste, H. (2010). Citizenship Education: A Critical Look at a Contested Field. En L.R. Sherrod, J. Torney-Purta & C.A. Flanagan (Eds.), *Handbook of Research on Civic Engagement in Youth* (pp. 161–188). New Jersey: John Wiley & Sons.

Hernandez, C.; Ravn, O.; Forero-Shelton, M. (2014) Challenges in a Physics Course: Introducing Student-Centred Activities for Increased Learning. *Journal of University Teaching & Learning Practice*, 11(2), Article 8, 2014. Available at: http://ro.uow.edu.au/jutlp/vol11/iss2/8

Höttecke, D.; Silva, C.C. (2011) Why Implementing History and Philosophy in School Science Education is a Challenge: An Analysis of Obstacles. *Science and Education*, 20, pp. 293–316.

Johnson, O.P. (1944) How can science education make its greatest contribution in the post-war period? *Science Education*, 28(4), pp. 231–231.

Longworth, N. (2003) Lifelong Learning in action Transforming Education in the 21st Century. New York: Routledge.

Mattos, C. R.; Tavares, L. B. (2014) The multiple senses of science teaching at a hospital school. European Science Education Research Association 2013 Conference e-Book Proceedings: Science Education Research For Evidence-based Teaching and Coherence in Learning. Nicosia: ESERA, v. 1. pp. 16–25.

Nathan, R. (2006) *My Freshman Year*: What a Professor Learned by Becoming a Student. New York: Penguin.

Novak, G, Patterson, E.T., Gavrin, A.D., and Christian, W. (1999) *Just-In-Time Teaching: Blending Active Learning with* Web *Technology*, Upper Saddle River, NJ: Prentice Hall.

Rodrigues, A; Camillo, J.; Mattos, C.R. (2014) Quasi-appropriation of dialectical materialism: a critical reading of Marxism in Vygotskian approaches to cultural studies in science education. *Cultural Studies of Science Education*, 9 (3), pp. 583–589

Schultz, T. (2001) Science Education Through the Eyes of a Physicist. National Academy of Sciences (2001). Available at: http://www.nas.edu/rise/backg2d.htm

Sundström, M.; Fernández, C. (2013) Citizenship education and diversity in liberal societies: Theory and policy in a comparative perspective. *Education, Citizenship and Social Justice*, 8 (2), pp. 103–117.

Tan-Wilson, A.; Stamp, N. (2015) College Students' Views of Work–Life Balance in STEM Research Careers: Addressing Negative Preconceptions. CBE—Life Sciences Education. 14 (3), pp. 1–13.

Vulliamy, G. (2010). Educational Reform in a Globalised Age: What is globalisation and how is it affecting Education world-wide?. *Early Childhood Education*, 1(5), pp. 1–16.

Wattar, D. (2014) Globalization, Curriculum Reform and Teacher Professional Development in Syria (Doctoral thesis) University of Alberta (Canada).

Wells, G. (2010). Schooling: The contested bridge between individual and society. *Pedagogies: An International Journal*, 5(1), pp. 37–48.

Welton, M. (2005) *Designing the Just Learning Society: A Critical Inquiry*. Leicester. UK: National Institute of Adult Continuing Education.

Wyer, M. (2003) The importance of field in understanding persistence among science and engineering majors. Journal of Women and Minorities in Science and Engineering. 9(3–4), pp. 273–286.

Chapter 9
Organizing Teaching to Solve Problems: The Case of Latitude and Longitude in Pre-service Primary Teachers' Education

Ruben Limiñana, Asuncion Menargues, and Sergio Rosa-Cintas

Abstract Most of the primary science school curricula across the world include the study of the Earth position in the universe, forgetting the problem of where we are on the Earth. Typically, contents related to location of people on the Earth begin with an already mapped and graded sphere on which geographical coordinates of a place are given as the angle between the place and the Equator and between the place and the Greenwich meridian. Hence, latitude and longitude appear as definitions, concepts already finished, without reference to the problem that is in the origin of their invention. In this way, there are not opportunities for engaging students in one of the problems whose historical solution has had an important practical (trade and navigation on the high seas) and conceptual (our place in the Earth and the Solar System) impact. Therefore, this could be an interesting issue for education of pre-service primary teachers, but only if our location on the Earth surface is stated as a problem. We approach this topic ("how can we know where we are on the Earth?") after students know that there are two days (equinoxes) in which the differences in the Sun path between different places are restricted only to the culmination value and time when it occurs. To address this problem, the professor's team had designed a sequence of activities, which has a structure that generates a tentative environment, where students and teacher have to plan a strategy to solve the problem, carry out this plan, and analyze results. In this chapter, we present the main activities of the sequence mentioned above, what students do, and the difficulties found along the problematic process that yields the invention of the so-called latitude and longitude.

Most of the primary and secondary science school curricula across the world include the study of the Earth position in the universe, forgetting the fundamental problem about where are we on the Earth (which was a historical problem for human beings). Typically, primary school contents related to location of people on the Earth begin with an already mapped and graded sphere on which geographical coordinates of a

R. Limiñana (✉) · A. Menargues · S. Rosa-Cintas
University of Alicante, Campus San Vicente del Raspeig, s/n, Alicante, Spain
e-mail: ruben.lm@ua.es

M. Pietrocola (ed.), *Upgrading Physics Education to Meet the Needs of Society*,
https://doi.org/10.1007/978-3-319-96163-7_9

place are given as the angle between that place and the Equator (latitude) and between the place and the Greenwich meridian (longitude). Hence, when people know those angles, they only have to "count" lines to locate a given place (or themselves) on the spherical Earth. Latitude and longitude appear then as definitions, concepts already finished, without reference to the problem that is in the origin of their invention: the location of human beings on the Earth. In that way, opportunities for engaging scholars in one of the problems whose historical solution had an important impact in both practical (trade and navigation on the high seas) and conceptual (our place on the Earth and in the Solar System) aspects, are lost.

Currently, there is a general interest in instructing future primary teachers in a way that enable them to be able to teach science in accordance with scientific practices (e.g., Lawson 2004; National Research Council 2012). In our case, we use this theme (our location on the Earth) as an opportunity to engage pre-service primary and secondary teachers on an inquiry-based science education (IBSE) methodology. In the case of pre-service primary teachers, it is especially important because they enter the university (at least in our faculty) having an overall prior education on humanities (80% of the students) and having mostly negative attitudes toward science teaching and learning. Only 25% of students have positive attitudes toward learning and teaching science, and they report that they have been taught during their secondary education using "a traditional" model (i.e., receiving the knowledge as it is, no tackling for the problem of interest on its origin) and, hence, they found most of these knowledge meaningless (as they could not use what they have learned to solve problems in their everyday life). Therefore, most of these students feel unconfident to teach science, as they have not the necessary scientific background (both conceptual and methodological) to do so. Attitudinal change and confidence to teach science using an IBSE methodology require that pre-service primary teachers have had opportunities during their formation for a "true learning" of any of the core or "big ideas" of science in a coherent manner with that we expect they follow in the future to teach their pupils (National Research Council 2012). Hence, our objective here is that students feel that they can learn any science topic in depth, which would result in an improvement in their attitudes, which is a key step for them to feel that they can also teach science with confidence.

The movement of the Sun and Earth is included in most of the primary education curricula across the world, and it may have an important practical interest, so this could be a good issue for the science education of pre-service primary teachers but only if location on the Earth surface is stated as a problem to be solved using scientific practices (according to our main aim). The episode we describe in this work is a subproblem belonging to a bigger one: "How should the Sun and the Earth move for the observed cycles and symmetries to occur? (the invention of a Sun/Earth model)." We approach this topic ("how can we know where we are on the Earth?") after students know that there are 2 days (equinoxes) in which the differences in the Sun path between different places on the Earth are restricted only to the culmination value and time of the day when it occurs. In this work we present the main activities of a more general sequence of activities, what students do, and the difficulties found along the problematic process that yields the invention of the so-called latitude and longitude concepts.

The methodology we encourage is an inquiry-based teaching model, but it is about fundamental problems of science (oriented research about fundamental problems of science) in this case about solving the question on where are we on a spherical Earth. That is to say, we do not aim to solve any question related to science, but the questions that are in the origin of the problem are the ones we intend to address. The main idea under this teaching methodology is that there are no fundamental concepts in science but fundamental problems to be solved that lead us to invent or develop some concepts to be of help in the resolution of that problem. In the present case, human beings had to invent the latitude and longitude concepts to know where they were on the Earth, which was possible after tracking Sun path and changes occurring on it over time.

9.1 How Is the Course Arranged?

As we have mentioned before, we develop this topic with pre-service primary teachers, which are expected to be generalist teachers in the future, in a subject during their training period at the university. The course consists of 60 h (4 h per week during 15 weeks) and we divide it in two parts. The first part (ca. 20 h) is addressed for students to express their initial attitudes about science teaching and learning, to become aware of the possible causes of their attitudes (among others, conventional teaching), and to realize about their willingness and confidence for teaching science at the primary school. This part of the course is also devoted to elaborate and justify a plausible model for teaching science in accordance with the scientific practices (as far as possible and desirable at the scholar level and characterized as a problem-solving and tentative activity). The second part of the course (ca. 40 h) is devoted to teach them the topic mentioned above, following the proposed model (i.e., teaching as oriented research about core problems and questions of science).

Within this topic, we set up two subproblems: a first more empirical subproblem (about cycles and symmetries in the Sun's movement as observed from a given place on Earth – "Are there changes and regularities in the movement of the Sun? Are these changes interrelated? Could we use the Sun's movement and its related changes for temporal and spatial orientation?") followed by a second more deductive (modeling) subproblem aimed at explaining empirical data observed when developing the first subproblem (about a Sun/Earth model that can explain the observed cycles and symmetries in Sun movement: "How do the Sun and the Earth move for the existing cycles and symmetries to occur?"). To approach this second problem, it is essential to put an observer on a spherical Earth in a given place, so this observer could exactly record the empirical data that we know that will occur in that given place. And to locate that place/observer on the Earth, we can only rely in data recorded from the Sun path in that place.

To address these problems, the professors' team had designed a sequence of activities that are proposed to pre-service teachers, which are organized in small groups

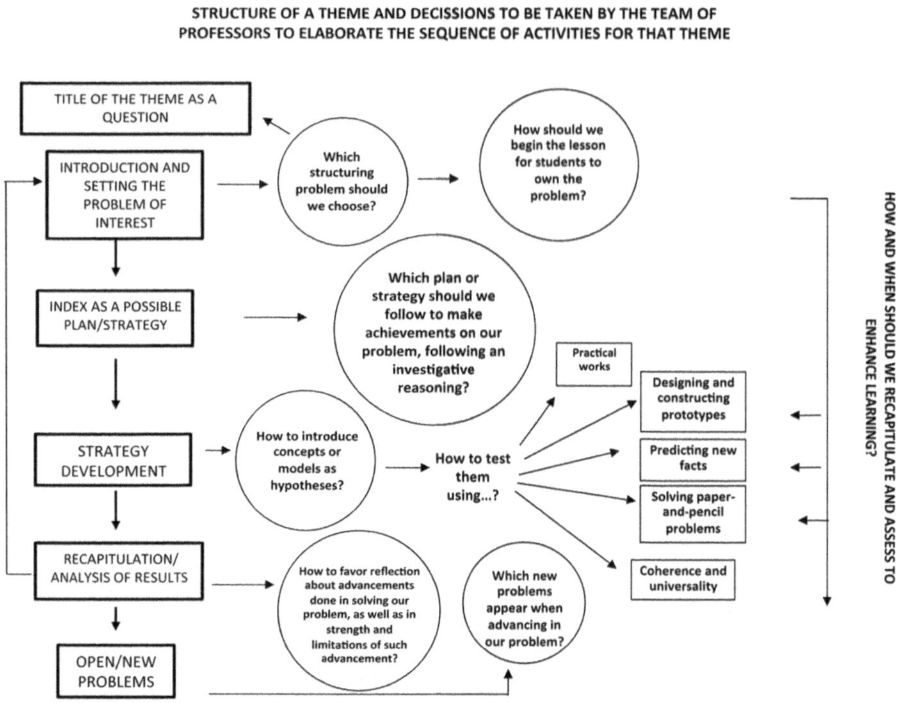

Fig. 9.1 General structure to elaborate a sequence of activities on a science topic using an oriented research on fundamental science idea approach

in the classroom. This sequence has a structure that generates a tentative environment (see Fig. 9.1) and serves to orientate and guide the students in the process of investigating about the problem of interest. In this environment, prospective primary teachers and the professor talk about what they have done in each activity, expressing their ideas about the problem that was set at the beginning, as well as their views and thoughts to every question in the sequence. For some activities, the professor provides empirical data (or reliable sources for students to find data) to help in doing the activities properly. Then, they assess, based on evidence and reasoning, if the ideas that have arisen as the result of responding to the activities in the sequence made a step forward to advance in the initial problem. When the professor explains or introduces some of the aspects or concepts, it is because it has been created for the need of his/her intervention, not just because it should be explained. This open and investigative atmosphere of early-stage researchers, guided by an expert (the professor), generates a climate that encourages the emotional involvement of pre-service teachers (Gil-Pérez and Carrascosa 1994). If students' ideas change, it is not because teaching has been planned to go against their initial ideas but because students take possession of explicit and scientific criteria to accept any idea, which can (or not) be different to "spontaneous knowledge."

When teaching science as oriented research, the plan to solve the initial problem of interest represents the index of the issue (see Fig. 9.2): the sequence of sections,

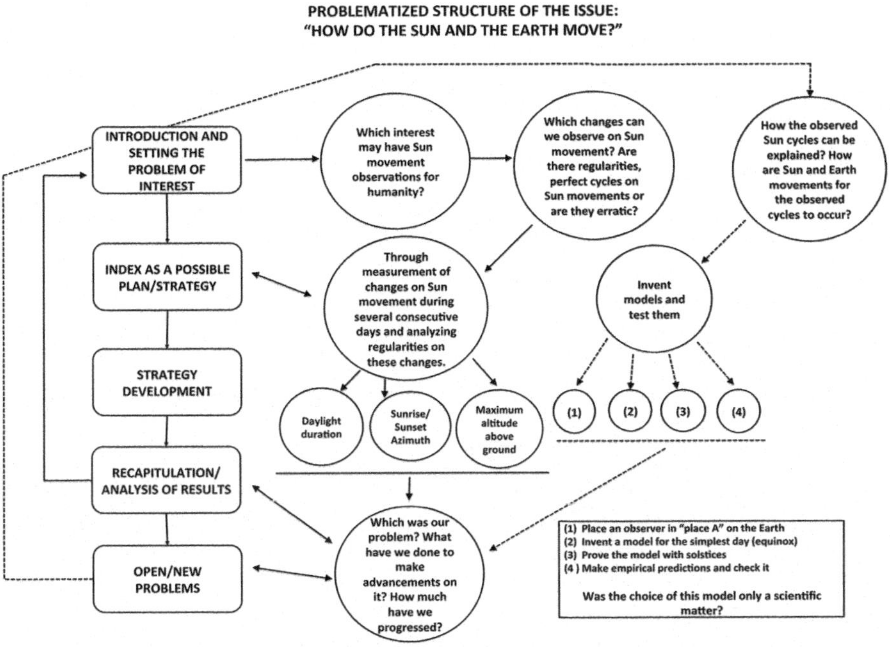

Fig. 9.2 Problematized structure of the issue taught in the course: "How do the Sun and the Earth move?"

steps, and activities is not arbitrary (for students), but it follows a logic structure (i.e., if this is the problem, what plan could we follow to make any advance toward its resolution?). Both students and professor should draw this plan (for students to be oriented during the research and for them to feel responsible of the plan). However, whether students cannot propose the entire plan and it must be completed by the professor, the plan proposal must be introduced as a tentative strategy and discussed with pre-service teachers for making it comprehensible and shared by them.

In the first problem, we begin with which variables could we use to describe Sun movement, arriving at the following: daytime duration, sunrise and sunset azimuths, and maximum Sun altitude in a day (culmination). Students have to make hypotheses on how they think that these variables vary over time. Then, data to contrast their hypotheses are provided. We first start with data from the place where they live, and, then, we can get other data to check what is the same and what is different from data collected from their city. During this process, we need to introduce some concepts, but, as stated before, these are only introduced when we have generated the need of having these concepts. For example, the concept of "cardinal points" is introduced when students are asked if there are changes in the sunrise and/or sunset position and if these changes occur regularly. Then, prospective primary teachers need to take a reference point in the horizon circumference to indicate the position of the Sun when it is on the horizon, and, at the same time, this point must be universal (they cannot use local references like mountains, buildings, trees, etc. if they need to communicate data to other people). Although students recall the compass and some of them explain why it allows to obtain a reference point on the horizon to measure the Sun's position from this point, the professor points out that before the compass was invented, there were accurate measurements of sunrise and sunset positions: the shortest shadow of a vertical stick (gnomon) occurs every day on the same straight line on the ground (i.e., the Meridian line). This allows organizing the horizon from the point where this line (in the sense opposite to the Sun) "intersects" with the horizon. In places where the Sun moves from left to right (if we are looking at noon), the intersection point on the horizon is called "north" (the reference point, 0°), and 360° of the horizon circumference is measured clockwise starting from the north. We can measure the position of the Sun on the horizon giving the angle between the straight line that links the observer's eyes with the north and the straight line that links the observer's eyes with the center of the Sun globe (when it rises or sets). This angle is called sunrise/sunset azimuth.

Overall, the main conclusion we have to arrive at is that there are cycles and symmetries in the observed Sun path during a long period of time, which lead to define what the year is, as well as the time periods within that year (seasons), according to "special" values of these variables. For example, regarding daytime duration, there is a day with the longest daytime duration within a year and 1 day when daytime duration is the shortest (these are the solstice days), and, similarly, there are 2 days in a year at which daytime duration is exactly 12 h (equinox days); defining and using those values and how daytime duration varies during a year enable us to define a year and seasons. The same applies if we use sunrise/sunset azimuths or culmination: we would observe the same pattern and the same existing cycles and symmetries; in fact, the three variables we have used are correlated (e.g., if we know the value of daytime duration, we can know which are the values of sunrise/sunset azimuth and culmination). And the most important thing is that we can use values of any of these variables (in the place where they live) for temporal orientation. And also, as we have described before, we can use Sun path for spatial orientation (see the example on cardinal points above).

Once we have studied variations in these variables that can be used to organize time and space, and how we could use these for our orientation, we can address the following problem: how can we use this knowledge to know where are we on a spherical Earth? (the development of a functional Sun/Earth model). To start, we need to recall that we know how Sun path is where we live (we already know the values of the aforementioned variables). Hence, we need to locate an observer in a sphere, so he can record the values of these variables; to do that, we have to deploy this observer on the Earth with all the necessary instruments to track Sun path. We also need to choose a day of the year to do this; it seems reasonable to use an equinox day, as daytime duration and sunrise/sunset azimuth that day are all the same at any place on the Earth, with the only difference between places being the culmination value. The change from being observers on the Earth to see that observer from outside the Earth is a fundamental step to understand how can we know where we are over a spherical Earth. Indeed, one of the most important activities in this second part of the course is deploying an observer on the Earth to track the Sun movements. This important step requires doing two important things: (1) place the Sun and the Earth in the position they are in an equinox day, and (2) find a place where the observer being there would record the culmination value we know to occur in that place in an equinox day.

Here, we present part of the sequence of activities designed by the teachers' team to solve the abovementioned problems. Complete sequence is available in the institutional page of our research group (Martínez-Torregrosa et al. 2016). As an example, the place, which this sequence refers to, is Alicante, Spain. Comments explaining and clarifying the activities are represented in italics.

Where Are We on the Earth? (Getting Oriented on the Spherical Earth)
If we want to develop a model to explain local observations about Sun movements, it is necessary that we identify in a spherical Earth some points (i.e., different places) that could correspond to places where the observer could be. To this end, we should first try to identify a place corresponding to Alicante in a day when the only difference between Alicante and other places on Earth is the different Sun's culmination value (i.e., in an equinox day). To begin, we will try to imagine how the Sun and the Earth must be placed for an equinox day to occur and, then, try to find a place where the observer would record a Sun's culmination value of 52°.

A.1 Make a proposal on how the Sun and the Earth should be located in an equinox day (i.e., for daytime duration being 12 h everywhere). If you can find different solutions, express how can we decide which is the most likely one. It is necessary that we could draw the Sun/Earth system clearly, so we can imagine what an observer on Earth sees. To make such a good representation, we are going to use a "lateral view" and a "zenithal view" (from above).

A.2 Draw the Sun/Earth system in an equinox day using the zenithal view (above the Earth's rotation axis, from quite far away) and lateral view (from a point in the perpendicular plane to the rotation axis, the plane that divides the Earth into two equal parts, also from quite far away). Once we have placed the Sun and the Earth for an equinox day to occur (and once we have determined in which direction the Earth rotates), we have to locate ourselves.

A.3 Place different observers in different places on the Earth with all the necessary instruments to track Sun path: at least a horizontal plane with cardinal points and a vertical gnomon. Try to imagine how these instruments look like if we were observing the spherical Earth from quite far away. Do this using a white foam sphere, and make the draws using the lateral and zenithal views (in the paper).

A.4 To locate Alicante, we know that Sun culmination (maximum angular altitude at noon) in an equinox day is 52°. Using the lateral view of the Earth, place three points on the Earth where Sun culmination should be different (higher or lower) to that in Alicante. *Using the lateral view, we need to locate an observer to record the Sun culmination value (to do that, that observer will*

(continued)

need to have a gnomon or vertical stick) in an equinox day in our city (i.e., 52° for Alicante, Spain).

Once this is done, how can we accurately determine which place on Earth corresponds to Alicante? (this question leads to introducing the concept of latitude). *Once we have located that observer in an equinox day measuring the corresponding Sun culmination value for this day, we have to bring up that the latitude of this person would be exactly 38° (the angle between the Equator and the place where that person is located). As all this discussion has been done in the blackboard, making the necessary drawings and representations to illustrate this, it is easy to realize then that culmination value in an equinox day in a place and the latitude of that place are complementary angles (i.e., they sum up to 90°). Hence, we can use Sun culmination value in an equinox day to know where we are on the Earth (i.e., our latitude); the angle that separates us from the Equator cannot be seen (as local observers on the Earth), but we can easily know what Sun culmination value in an equinox day is and, hence, use it to know where we are. Latitude values range between 0° (Equator) and 90°S (South Pole) or 90°N (North Pole).*

As we have already seen, only the latitude is not enough to determine where Alicante is, given that all the places at the same latitude (i.e., being in the same parallel) would measure the same Sun culmination in an equinox day (indeed, all days of the year). The only difference related to Sun path between all those places is the time of the day (i.e., the hour) when the Sun reaches the maximum angular altitude above ground (the local noon), if we were using the same clock to record that hour in all those places. Therefore, if we had a "universal clock" for us to compare when noon occurs in different places, we could then be able to properly identify a given site by using its latitude and time of the day when noon takes place.

A.5 We already know how to determine the latitude of any place on Earth, but this is not enough to locate that place on the spherical Earth. How can we properly locate that place? Read the following paragraphs until you know how could you determine the geographical coordinates (latitude and longitude) of a given place on Earth.

All places on Earth where the Sun reaches the maximum angular altitude at the same time of the day represent an imaginary line named meridian (draw several meridians using a zenithal view of the Earth). If we could use synchronized clocks everywhere (a "universal time") and take one of the meridians as a reference (that "passing" by the Greenwich observatory and very close from Alicante), we can calculate the angle defined by the Greenwich meridian and that passing for any place on Earth comparing the time interval when noon occurs between these two places. We can do this knowing that any point on Earth lasts 24 h in being again at the "same place" (i.e., to complete a turn of 360°). Hence, in an hour the Earth rotates 360°/24 h = 15°. If noon in a

(continued)

given place takes place 2 h later than in Greenwich, it means that place is 30° away from Greenwich. And the angle between Greenwich meridian and the meridian of any place is called the longitude of that site. Places located eastward from Greenwich would see noon before it is seen in Greenwich, and those located westward from Greenwich would see noon after it is seen in Greenwich (as a consequence of the direction of the Earth's rotation). We could thus know that longitude ranges between 0° and 180°E and 0° and 180°W. Here, we have mentioned the importance of having clocks to measure a universal time for us to be oriented in the spherical Earth. The invention of accurate clocks to be used in open seas was necessary for navigation and, hence, for commerce and economical development (Fig. 9.3).

Once we have located an observer where we live, we complete the course studying if the Sun/Earth model developed for equinox days explains what we know to occur in those days (Sun rising in the east and setting in the west and daytime duration is exactly 12 h) and what should happen for us to record the values of culmination (and the other related variables mentioned above) in solstice days. We are not developing this part in this work.

9.2 What Can Students Do After Instruction?

After instruction, students can solve problems related to temporal and spatial orientation, as well as problems related to location of people on the Earth, using only data related to Sun movement and its pathway. Students are also required to provide reasoning and justification of their calculations. We assess the content

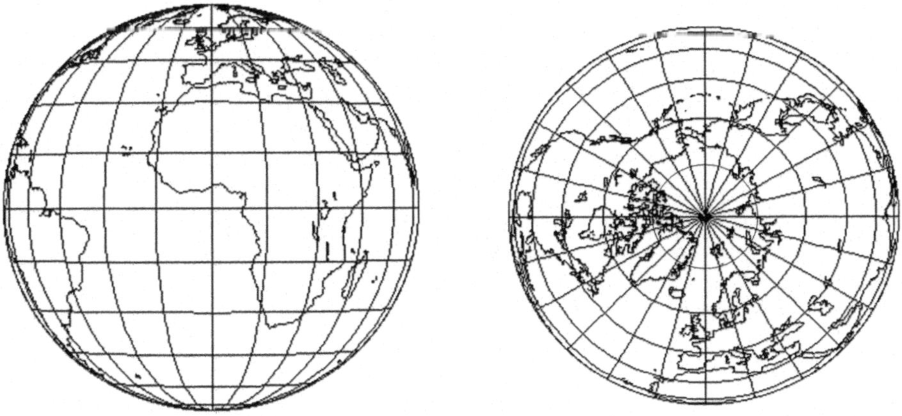

Fig. 9.3 Meridians and parallels using a lateral and zenithal view of the Earth

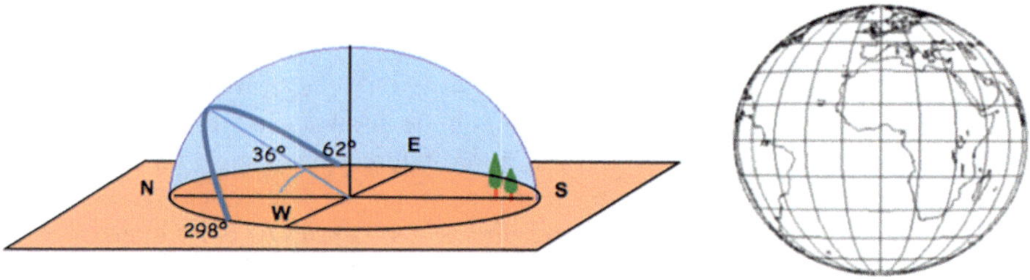

Fig. 9.4 Question about the Sun path used to assess knowledge level

knowledge achievement through questions in exams. Here are some examples of questions that are used to assess the knowledge level achieved by prospective primary teachers after the instructions as oriented research in this topic:

– The following drawing (Fig. 9.4) shows the Sun path in the winter solstice in an unknown place on the Earth. Moreover, the person who has made the drawing has told us that gnomon shadow is on the Meridian line two hours after than in Greenwich. Could you locate that place on the Earth?
– Please use the Sun/Earth model to deduce how Sun path will be in the winter solstice in a place whose latitude is 55°N and longitude is 90°W. Please compare maximum Sun altitude (culmination), daytime duration and sunrise and sunset azimuths with data from Alicante.

9.3 Conclusions

In this work, we have shown how this science topic is taught to pre-service primary teachers using an oriented research approach. As we have mentioned before, questions to be solved using this approach are those that are in the origin of the problem of science that we want to address. That is to say, research is not done on "accessory" topics but on fundamental ones. In this case, the fundamental problem is the development of a Sun/Earth model, which implies knowing where are we on a spherical Earth.

The methodological approach has resulted in students reaching a high content knowledge on the topic. Prior to beginning the course, we used a questionnaire written in everyday life words to assess the initial knowledge of these prospective teachers. Those results indicated that initial knowledge was usually low and plenty of conceptual mistakes on the topic (Martínez-Torregrosa et al. 2018). However, after instruction ca. 80% of pre-service primary teachers achieved the minimum expected content knowledge, thus being able to use the Sun/Earth model to solve questions as those described above.

References

Gil-Pérez, D., & Carrascosa, J. (1994). Bringing pupils' learning closer to a scientific construction of knowledge: a permanent feature in innovations in science teaching. Science Education, 78, 301–315.

Lawson, A. E. (2004). Biology: an Inquiry Approach. Dubuque, Iowa: Kendall/Hunt Publishers.

Martínez Torregrosa, J., Martínez Sebastiá, B., Osuna, L., Menargues. A., Limiñana, R., Colomer, R., Savall, F. & Rosa, S. 2016. *How do Sun and Earth move for the existing cycles and symmetries to occur? (The invention of a Sun/Earth model).* Available at: https://rua.ua.es/dspace/handle/10045/2601

Martínez-Torregrosa, J., Limiñana, R., Menargues, A., & Colomer, R. (2018). In-depth teaching as oriented-research about seasons and the sun/earth model: effects on content knowledge attained by pre-service primary teachers. Journal of Baltic Science Education, 17(1), 97-119.

National Research Council. (2012). *A framework for k-12 science education: Practices, crosscutting concepts, and core ideas.* Washington, DC: The National Academies Press

Chapter 10
Innovation in Physics Teaching/Learning for the Formative Success in Introductory Physics for Bio Area Degrees: The Case of Fluids

Marisa Michelini and Alberto Stefanel

Abstract Physics course for student in bio area scientific degrees is a multidimensional problem, where the acquisition by students of a functional understanding of physical concepts is the main problem. To face this problem requires a strong revision of topics and of approaches to physics concepts. Physics has to be problematized and offered to students as a useful tool for their future study and job in contexts which are related to the bio area. Design-based research intervention modules were studied in the last two years, taking into account the above-mentioned needs, for degrees at the University of Udine of Agricultural Science and Technology, Biotechnology, Environmental and Nature Sciences and Technology, Oenology, and Science of foods. The courses involved two cohorts (2014/15 and 2015/16), respectively, of 342 and 483 students. Each course covers the classical physics and consists of three modules. Physics of fluids is here selected as topic characterizing in different ways the professional education of student in such degrees. The characteristic of the intervention module on fluids exemplifies the approach followed, testing the effectiveness and documenting the students' learning outcomes.

A positive general trend emerges in the average students' learning outcomes, indicating the effectiveness of the proposal implemented. The more problematic aspects for 10–30% of students are related to the concepts of pressure and the Pascal principle, the bridge from static to dynamic situations. The engagement of students in analyzing those questions that are typically evidenced as learning problems is effective not only in overcoming the single specific aspects but also in facing new situations. The management of math for students is critical as well as the confidence with the validity range of a physical law.

M. Michelini · A. Stefanel (✉)
Research Unit in Physics Education, Department of Mathematics, Informatics, Physics, University of Udine, Udine, Italy
e-mail: alberto.stefanel@uniud.it

M. Pietrocola (ed.), *Upgrading Physics Education to Meet the Needs of Society*,
https://doi.org/10.1007/978-3-319-96163-7_10

10.1 Introduction

Physics course for student in scientific degrees in the bio areas is a multidimensional problem, mainly concerning what Lillian McDermott defines a functional understanding of physical concepts (McDermott and Shaffer 1992; McDermott et al. 2006). That is, students must be able to apply the physics concepts in the different contexts of their specific field of study and their professionalism. The main aspects to be faced are:

(A) To redesign the way in which physics is offered so that its role can be recognized in the specific subject matter characterizing the degree: turning the ways in which physics is approached, changing the role of each topical areas, and individuating specific applications of physics in the professional field of the degree (Cummings et al. 2004; Hoskinson et al. 2014; Meredith and Redish 2013; O'Shea et al. 2013)
(B) To offer instruments and methods building a physics competence in different fields (Hoskinson et al. 2013)
(C) To individuate strategies able to produce an active role of students in learning physics and to give them the opportunity for an appropriation of the applied physics methodologies (Hoskinson et al. 2013)
(D) To support student learning in multitasking ways by means of ICT tools, of lab activities, of problem-solving, and of step-by-step evaluation of learning outcomes (Laws 2004; Redish and Hammer 2009; Meredith and Redish 2013)

Design research-based intervention modules (Collins et al. 2004) on physics were studied in the last 2 years, taking into account the abovementioned aspects, for degrees in the University of Udine of C1, Agricultural Science and Technology; C2, Environmental and Nature Sciences and Technology; C3, Oenology; C4, Science of foods; and C5, Biotechnology (Michelini and Stefanel 2016). The courses involved two cohorts (2014/2015 and 2015/2016), respectively, of 342 and 483 students. Each course covers the classical physics and consists of three modules. The physics module for the bio area has a common heart and was differentiated concerning the context-related activities (exercises, applications considered). Physics of fluids is here selected as topic characterizing in different ways the professional education of student in such degrees.

The characteristic of the intervention module on fluids is discussed together with the student learning outcomes.

10.2 Theoretical Background

The innovation in introductory and upper-level STEM education is actually a challenge including both disciplinary and interdisciplinary efforts in established and emerging fields, targeting all institutions of higher education and involving research groups and associations of different subjects (AAAS 2004). Physics education research focused on the innovation of physics course for Undergraduate Science Curriculum identifying, developing, adapting, implementing, disseminating, and assessing of exemplary educational materials (tutorial), processes, and models for active learning (Heron et al. 2004; Laws 2004). These research also became examples for developing new curricula for physics course in other areas as, for instance, the biology one (Donovan et al. 2013), as well as to integrate current science into the physics curriculum ad in general to promote the student learning in interdisciplinary biology and physics classes (AAAS 2011, p. 54; Brewe et al. 2013). The first basic question is how to integrate physics into biology curriculum (or biology into physics curriculum) beyond simple provision of examples from the respective disciplines (CBE 2013). The request of interdisciplinarity opens new research questions concerning starting assumptions about students, content to treat, competencies to focus on, corridors and barriers to constructing an effective course, and condition and resource for effective inter- or transdisciplinary instruction (Redish et al. 2014). Future research aims could be how sophisticated, biologically relevant physics topics can be taught at the introductory level and how biology instruction should be changed such that students are prepared to use physics knowledge and theory to understand biological phenomena. Adding problems emerges from the different perspectives and epistemologies of physicists and biologists. Topics that physicists view as "canonical" and considered important for all students are quite different from topics that biologists view as important for understanding and doing modern biology, as, for instance, random motion, diffusion, microstate thermodynamics, and fluid flow (CBE 2013; Redish et al. 2014).

Physicists usually isolate the object of study to be able to focus on fundamental processes in systems with a (relatively) small number of degrees of freedom. In biology, the systems are always interconnected and not separable. Moreover, essentially everything takes place in a fluid environment—air or water—and the fluid has a critical influence on biological function. For that the dynamic of fluid is essential to include in a physics course for bio area, and other topics can or must be eliminated, as, for example, the projectile motion (as paradigmatic example) or gravity (Meredith and Redish 2013).

In this vast area of concerns, some studies addressed the role for an integrate learning of methodological aspects, such as problem-solving and modelling (Hoskinson et al. 2013, 2014) or crosscutting themes as, for instance, energy (Cooper and Klymkowsky 2013; Svoboda Gouvea et al. 2013; Dreyfus et al. 2014), usually proposed in very different ways, for instance, in physics courses and in biology courses, producing fragmented understanding (Svoboda Gouvea et al. 2013) or contradictory, inconsistent conceptions in the students (Dreyfus et al. 2014). Other scholars studied how physics can be used to explain significant biological processes and phenomena (Bustamante 2004), the possibility to give a formal description to biological phenomena (Redish and Cooke 2013). Although the role of math in biology increases, some research evidenced that students do not have the same perception (Hall et al. 2011, Watkins and Elby 2013). The role of integration in producing capacity to integrate knowledge and modes of thinking in two or more disciplines was also explored in different dimensions (Ivanitskaya et al.

2002; Boix et al. 2007). General criteria are proposed for the evaluation of new learning objectives, integrating physics, and biological thinking (Watkins et al. 2012; Svoboda Gouvea et al. 2013, Thompson et al. 2013). Recently a great effort was made to design new physics courses entirely curved on biology (Cummings et al. 2004; Meredith Redish 2013; O'Shea et al. 2013), adopting innovative curriculum models (Watkins et al. 2012; Manthey and Brewe 2013; Donovan et al. 2013, Thompson et al. 2013) to help students develop reasoning strategies that move beyond traditional disciplinary boundaries.

As discussed by Svoboda Gouvea et al. (2013), almost three are the levels of integration of physics and biology tasks:

1. Features of Level 1 Tasks—Superficial Interaction describes a relatively low-level interaction between disciplines, (as, for instance, the fish buoyancy problem, where biology is just a context where students are requested to reason about the Archimedes law and the buoyant force acting on a fish).
2. Features of Level 2 Tasks—One Discipline Impacts the Other, one discipline impacts or modifies a second in some substantial way (e.g., this might mean applying a technique that is common in physics (e.g., dimensional analysis) or in the case when physics law explain a bio aspects (as pressure to understand the arteriosclerosis formation).
3. Features of Level 3 Tasks—Exploring Connections Between the Disciplines, where the integration occurs, for instance, bringing different conceptual frameworks of each discipline to bear on a problem and explicitly examining why these frameworks differ and where they overlap (as in exemplum analyzing ATP cell role with an energy perspective).

Our work goes in the perspective to give a contribution on teaching physics to bio area in an integrated way according to above level 2–3 tasks. We follow the Redish group suggestions, in particular concerning the need to change radically the approaches, rooting the treatment of physics content in context interesting for bio area, focusing more on fluids and fluid dynamic more than to the mechanics of material points (Meredith and Redish 2013; Cummings et al. 2004; Redish et al. 2014). We stress also the importance to include experimental lab, interactive lectures, and high student engagement (Redish and Hammer 2009). We discuss the general characteristic of our approach, exemplifying it in the case of the module of fluids and presenting student learning outcomes.

10.3 The Research Questions

The present contribution focuses on the following research questions: RQ1—Which contents are more problematic for student learning? RQ2—Which role does play the engagement of students in the analysis of questions typically evidenced as learning problems? RQ3—Which kind of reasoning evidence students facing these questions?

General Characteristics of the Course in Physics for Bio

As indicated before, following Redish approach (Cummings et al. 2004; Meredith and Redish 2013), physics is offered usually starting from context and applications of bio areas. For instance, a contextualized problem-solving introduces the motion issues: "A cheetah hunting an antelope: will the cheetah reach the antelope?". The usual encountered problem becomes a problem-solving starting from the characteristic of the involved animals (the maximum speeds, the typical acceleration, the distance covered with maximum speed). Examples are taken from the environment, wine production, and food preparation, focalizing where physics is important to understand a "bio" phenomenon. For instance, fluid dynamics is treated not only in ideal conditions but also in the water flow in an open and closed duct, in the river flow, in the blood circulation, and in the respiratory apparatus in human body.

A privileged attention is given to the fluid dynamics, compared to those given in the traditional physics courses, where the prevalence is on fluid statics. As concern strategies, we adopt a context-related problematic approach to each topic using sometimes flipped classroom strategies to engage students in finding the general physical behavior that lies at the base of the problem proposed. As concern methods interactive lecture demonstrations are integrated with group work problem-solving activities, analysis of contextualized problems (problem facing situations typical of the science area), seminars, and labwork. Frequent formative learning outcomes evaluations are proposed by means of clickers questionnaires and traditional multiple-choice or open questionnaires and only for C5 students open problems. After each learning outcomes evaluation, an in-depth discussion is carried out with student on learning knots emerged.

Usually each hour of the courses is organized as follows: 45 min of lessons, using blackboard, PP presentation, and demonstrative experiments, and 15 min of clickers or manual clicker-like sections, exercise, and simple applications.

The Module on Physics of Fluids

Table 10.1 reports the schema of the module, indicating the main contents included and the number of hours dedicated.

On the physics point of view, the fluids are introduced as systems flowing, not reacting to transverse forces (static) or F//surface = 0. Their description requires a mesoscopic modelling and a change in the typical quantities used (density and pressure vs. mass and forces in the Newtonian dynamic of material points). The concept of pressure is discussed in three perspectives: as a force distributed on a surface, as (normal) a compression on the surface of the fluid, and as a state property of a fluid. In the third perspective, the Pascal principle emerges as the general law characterizing the specific behavior of fluids in equilibrium or near the equilibrium and distinguishing with respect to the solid systems. Stevin and Archimedes laws are

Table 10.1 Schema of the module on fluid

Hours	Contents
2	Physics of fluids in equilibrium (fluid as a continue system that can flow, pressure concept and Pascal Principle, Stevin and Archimedes laws, density concept and its role in buoyancy)
2	Dynamics of fluids (flux and flux conservation equation, Bernoulli theorem)
2	Real (more realistic) fluids (the concept of viscosity, Stokes' force, Poiseuille equation, capillarity, surface tension)
1	Problem and exercises
2	Lab (experimental study of balls falling inside different liquids—only in the two courses AGNV and STF courses)

discussed in many important applications and examples, as for instance: the communicating vessels; the dam; the measurement of χ in liquids with the piezometer; the Mariana trench and the gradient of pressure with depth in a liquid; the hydraulic torque; the U-tube manometer; the syphon; the Torricelli and other barometers; the heart pressure in human and giraffe; the hydrostatic paradox; the density meter, discussed also as a rigid rotating body, and the alcohol meter; the buoyancy of liquid in liquid; the Archimedes forces in the air as in the candle flame; the Montgolfier; the bulb and the cylinder on the equal arms balance and the vacuum pump; the measurement of the hydrostatic force; the King Hiero legend. The surface tension, introduced as work to enlarge a liquid surface, and the surface phenomena are introduced as important aspects of physics of fluids in living systems, starting from capillarity in the trees as well in the peripheral blood circulation. Laplace formula and Borelli-Jurin law are discussed and applied to explain that phenomenology as well as specific examples as the drop method to measure the tension coefficient, or the emboli formation.

The dynamic of fluids is discussed starting from the phenomenology of real cases (i.e., the water flow in a river, as well as the blood circulation in the body). Continuity equation is stated as the general condition that must be satisfied by a fluid flow. The complex case of turbulence and viscosity motion is discussed qualitatively, as base for a quantitative formal approach with simplifying assumptions, that are valid in specific situations concerning the context of applicability of Poiseuille law and Bernoulli theorem.

As claimed before, and more extensively discussed in a previous work (Michelini and Stefanel 2016), the water flux in a river is a typical introductory context to discuss the dynamics of fluid. The approach starts from the video analysis of the motion of the water in a real river at different distances from the riverside. The parabolic profile of velocity is then discovered analyzing the water flow in a rectangular duct in the physics lab and then modelled assuming that friction forces acting between contiguous layers of flow are moving at different speeds. The viscous forces and the layer models emerge as consequences of experimental observation and not as a priori assumption or abstract hypothesis. The Poiseuille law is also contextualized in the blood flow, where the typical range of validity is well satisfied. The continuity equation and Bernoulli theorem are discussed to interpret the formation of a plaque in an artery stenosis and atherosclerosis on the base of the pressure concept.

The analysis of blood circulation given to us to stress another point characterizing our approach on physics for bio. That analysis, in fact, shows that physics underlying blood circulation is the same as that of other physical systems, as, for instance, that of electric circuits. The analogy between blood circulation and electric current flow is based on an effective correspondence between elements (heart-battery; close blood system-close electric circuits; viscous resistance elements-electric resistors) and concepts (differences of pressure and differences of electric tension as driving factors, continuity equation, and charge conservation). The analysis of the decreasing of the arterial blood pressure from the hearts to the peripheral areas, interesting in itself, becomes a powerful context which activates the analysis of electric circuits on the base of decreasing of the potential along an electric circuit and vice versa.

This kind of approach activates students' model-based reasoning (Nersessian 2002) giving the opportunity to construct competencies on physics concepts contextualized in their own field and an integrated vision of physics models as useful model to describe different phenomenologies both important for bio area and for physics.

10.4 Instruments, Method, and Contexts of Evaluation

To evaluate the effectiveness of our approach and in particular to answer our research questions, we analyzed the questions submitted to the students for their (written) examination. Appendix reports the collection of question concerning fluid here considered. The format reported is that of multiple-choice questions, but in the different courses, the format of the questions contained some little change. In particular, in the course BT1–2 an explanation was explicitly requested and evaluated; in the courses AGNE1 and STF1, students get motivated of the choice done, also when not explicitly requested. In the courses AGNE2 and STF2, questions 1, 2, and 4 were proposed as open questions in intermediate questionnaires proposed during the lessons to the students.

The question proposed are of two types: qualitative questions on fluids in equilibrium concerning knots typically evidenced as learning problems (Loverude et al. 2010), to evaluate the conceptual understanding of students, and quantitative questions aiming to test the functional understanding of basic concepts of fluid dynamics.

The sample here considered consists of two cohorts of students attending the courses in 2014–2015 and 2015–2016 composed, respectively, by 342 and 483 students per each academic year as detailed in Table 10.2. Only the C5 are selected students on the base of a test with the same criteria at national level, producing an admission of 60%. The C1–C4 students are not selected students, only half of them with a modest preparation in physics and the other half with no previous preparation in physics at all.

Table 10.2 Composition of the sample of the two cohorts of students (AY 2014–2015 and AY 2015–2016) in the three courses concerning five degrees

Course	Degree	AY 2014–2015	AY 2015–16
Course 1 AGNE	C1-Agricultural Science and Technology	$N_{AGE1} = 186$	$N_{AGE2} = 261$
	C2-Environmental and Nature Sciences		
	C3-Oenology		
Course 2 STF	C4-Science of foods	$N_{STF1} = 110$	$N_{STF2} = 177$
Course 3 BT	C5-Biotechnology	$N_{BT1} = 46$	$N_{BT2} = 45$

10.5 Methodology of Data Analysis

A quantitative analysis of answers given by students to the multiple-choice questions was performed to extract indication about the general outcomes of our educational approach, especially with the signs of which aspects were better learned and what difficulties persist.

Using the qualitative research criteria, it also carried out an analysis of the student patterns of resolution and of reasoning expressed in the motivation/explanation of the choice of students.

Analysis of Student Learning Outcomes and Reasonings

Table 10.3 resumes data concerning Q1–Q8 items. Concerning Q1 the mean percentage of correct answers is 43% for the two cohorts, 47% for cohort 1 and 35% for cohort 2. The difference is explained by the different modes of administration of the question: closed questions in the first case and open-ended questions for the AGNE2 and SFT2 cohorts. In the few explanations, the prevalent strategy of solution was the dimensional analysis of the equation proposed (70% of student explaining the choice). Students showed difficulties in the individuation of the physical dimensions of χ, in the inversion of the formula that defines the compressibility coefficient χ, and in the passage from DV to Dρ. Therefore, it is not a surprise that the percentage of correct answers collapse to 13% and 24% in the case of AGE2 and STF2.

The percentage of correct answers to question Q2 is 68% for the overall sample, 78% for cohort 1, and 47% for cohort 2. Also in this case, the main difficulties are related to the inversion of the formula used (that defining pressure). Therefore, the better results in Q2 with respect to Q1 seems more related to the numerical format of the question Q2, than to other aspects. (Table 10.3).

Fifty-eight percent of the overall sample gave the expected answer to Q3 (equal pressure at the same level in the two arms of the container), without differences between the two cohorts. Usually students explained the answers referring to the Stevin law (because of "Stevin law") and/or the fact that "points at equal level have

Table 10.3 Percentage of responses to questions Q1–Q8 (see Appendix).

		BT1	BT2	AGNE1	AGNE2	STF1	STF2	TOT1	TOT2	TOT
		N = 46	N = 45	N = 167	N = 41	N = 108	N = 68	N = 321	N = 154	N = 475
Q1	**A**	**9**	**27**	**53**	**13***	**55**	**24***	**47**	**35**	**43**
	B	35	16	27	7*	22	6*	26	9	21
	C		11	20		15	7*	15	28	19
	NA	57	47		80*	8	63*	11	28	16
Q2	**A**	**67**	**76**	**80**	**27***	**79**	**40***	**78**	**47**	**68**
	B	22	2	17	51*	18	47*	18	35	24
	C	2	9	3	20*	3	12*	3	13	6
	NA	9	13		2*	1	1*	2	5	3
		N = 46	N = 45	N = 158	N = 41		N = 68	N = 204	N = 154	N = 358
Q3	**A**	**70**	**64**	**53**	**61**		**54**	**57**	**59**	**58**
	B	20	20	29	10		28	27	21	24
	C	4	4	18	5		13	15	8	12
	NA	7	11		24		4	1	12	6
		N = 46	N = 45	N = 158	N = 218	N = 99	N = 120	N = 303	N = 383	N = 686
Q4	**A**	**63**	**27**	**47**	**62**	**73**	**57**	**58**	**56**	**57**
	B	4	4	18	9	13	28	14	14	14
	C	4	9	29	22	8	15	18	18	18
	NA	28	60	6	7	6	1	10	11	10
		N = 46	N = 45	N = 120		N = 92	N = 120	N = 258	N = 165	N = 371
Q5	**A**	**37**	**31**	**49**		**77**	**38**	**57**	**35**	**70**
	B	17	2	23		14	18	19	12	23
	C	7	7	24		8		15	3	16
	NA	39	60	3		1	44	9	50	30
		N = 46	N = 45	N = 120		N = 92		N = 258	N = 45	N = 303

(continued)

Table 10.3 (continued)

		BT1	BT2	AGNE1	AGNE2	STF1	STF2	TOT1	TOT2	TOT
		N = 46	N = 45	N = 167	N = 41	N = 108	N = 68	N = 321	N = 154	N = 475
Q6	**A**	**76**	**49**	**73**		**64**		**71**	**49**	**67**
	B	4	9	5		0		3	9	4
	C	2	13	21		35		22	13	21
	NA	17	29	1		1		4	29	8
		N = 46	N = 45	N = 120	N = 41	N = 92	N = 68	N = 258	N = 154	N = 412
Q7	**A**	**67**	**47**	**63**	**26**	**78**	**68**	**69**	**51**	**62**
	B	2	18	29	74	14	32	19	39	26
	C	2	4	7		5		5	1	4
	NA	28	31	2		2		7	9	8
		N = 46		N = 120		N = 99		N = 258		N = 258
Q8	**A**	**43**		**72**		**80**		**70**		**70**
	B	2		8		10		8		8
	C	9		18		9		13		13
	NA	46		2		1		9		9

The answer (A) is the (more) correct one (in bold% of correct answers). (*: submitted as open question). In Q3–Q4–Q5 the percentage of the total sample are related to the number of students

equal pressure." Students motivated B answers evidencing three different ways of reasoning: the first is based on the liquid level "above the head" ("...the point K presents a mass of water over it greater than J"); the second stresses the role of the atmospheric pressure on the open arm ("in K, also Po is acting"); and the third starts from the definition of pressure and motivates the different pressures with the different sections of the two arms. The third way of reasoning is based on an arbitrary assumption that the same force is acting on the two arms and the wrong use of proportionality. The same assumption and the correct use of proportionality motivate some C answer. Another motivation for answer C considers what happens if "I open right arm.... → PK > PJ." In this case, the student evidences the idea that the pressure remains the same in opening the right arm.

Questions 4 and 5 regard the analysis of the dynamic of fluids. More than half of students (57%) performed the numerical evaluation requests in question 4. The main strategies of solution are the following: the combination of continuity equation and the Bernoulli principle to perform the computation arriving to the results and a qualitative reasoning, underlying the same laws—S decreases → v increases → P decreases. The answers B and C usually are motivated forcing the manipulation of the same equations to give the expected answer. The majority of students (70%) performed the requested analysis of dependence of parameters in Q5. As in the previous case, the main strategies of solution include an explicit manipulation of the Bernoulli equation and continuity principle. B answers are motivated referring to the Stevin law (P equal at equal h) or to the Pascal principle (P equal in any points). Concerning Q6, 67% of the full sample individuated on the velocity profile the level at which the flow speed is half of that of the superficial layer. The main reasoning is based on the linear relation between v and h. The percentage of answers changes proposing different profiles of velocity (i.e., a quadratic profile), evidencing the needs to go behind the use of the linear proportionality.

In the analysis of the Venturi tube, proposed in Q7, 62% of students correlated correctly the pressure and the level of the liquid in the corresponding column in each section of the tube. A higher percentage was observed for the groups performing clicker/clicker-like sessions. The main reasons for the answer (A) are based on the Bernoulli theorem, reconstructing qualitatively the chain: P decrease, decreasing the section. Other explanations are tautology ("PA > PB because the flow exerts a bigger pressure in A than in B") or descriptive ("the left arm push on the right one").

The velocity behavior of a ball falling in a liquid (question Q8) was individuated by 70% of students. The AGNE and STF students explored the phenomenon in the lab that seems at the base of better results. The main argument is based on the idea that the falling ball reaches a regime/limit speed. The option C is based on the expectation of an exponential trend. Option B is based on the application of a proportional reasoning.

10.6 Results and Conclusion

A study was performed on physics teaching-learning in degrees of the bio area, having as a framework the researches carried out in that field and in particular referring to the approach and results of the studies of the Redish groups. The main characteristic of the courses designed was here presented, focalizing on the module on fluids. The outcomes of students learning here discussed emerged from the analysis of the question presented in appendix and regarding the answers to our research questions.

A positive general trend emerges in the average student learning outcomes, indicating the effectiveness of the proposals. At the same time, we can discuss the more controversial aspects emerged by the analysis. First, the more problematic aspects for 10–30% of students regard the concepts of pressure and the Pascal principle. We observe also that some open knots remain passing from static to dynamic situations, more than considering the dynamic cases itself. For the students of the bio area, there is a critical management of math, as, for instance, in the analysis of the inverse problems and in going over the use of the proportional reasoning. Students of our sample show a weak preparation in math mixed with the epistemological obstacles observed in literature (Watkins and Elby 2013). Moreover, 20–25% gave no answers. This indicates the existence of a possible threshold in the construction of a functional understanding of physical concepts (RQ1).

Concerning RQ2, we have an indication that the engagement of students in analyzing questions typically evidenced as learning problems is effective not only to overcome that specific aspect but also to face dynamical situation. A threshold seems to exist also in this case. In any case a link between qualitatively/conceptual and quantitative questions emerged by data.

In the answers of quantitative questions, the majority of students adopted a (direct) proportional reasoning. This is for many students the only formal source that they adopt. It seems that the proportionality is the construct which builds the formal thinking of these students, and a great effort must be done to go over this construct in itself. Students used a descriptive/qualitative approach as a tool for prediction, more frequently than the interpretative/quantitative way of reasoning. About 20% of the student showed the extension of validity range of a physical law was out its range of validity (RQ3).

Our future research work will address how to increase the level of integration of physics in the areas of biological sciences in our approach, studying in particular how to improve the competencies of the students in the use of formal/mathematical tools.

Appendix: Items Included in the Written Questionnaires

NB: The first answer is considered the (more) correct one. A random order was used submitting the questions to students.

Q1. Water at environment pressure and temperature have a density of $\rho = 1000$ kg m^{-3} and a compressibility of $\chi = 6{\cdot}10^{-10}$ Pa^{-1}. Which expression give the variation $\Delta\rho$ of the density of water when is pressure increase of ΔP?

(A) $\Delta\rho = \rho\chi\Delta P$ (B) $\Delta\rho = \rho\chi/\Delta P$ (C) $\Delta\rho = \Delta P\chi/\rho$

Q2. A submarine is located at a depth such that the pressure exerted by the water on its walls is equal to $2.5{\cdot}10^5$ Pa. The portholes of the submarine have circular flat surface whose area is 0.03 m^2. What is the intensity of the resultant force with which the water pushes a porthole toward the interior of the submarine?

(A) 7.510^3 N (B) 8310^6 N (C) 4510^3N

Q3. (Elaboration from Loverude et al. 2010) A container, such as that shown in the figure, is formed by the left open branch and the right closed branch. Compare the pressures at points J and K. Which relation is correct?

(A) $P_J = P_K$ (B) $P_J < P_K$ (C) $P_J > P_K$

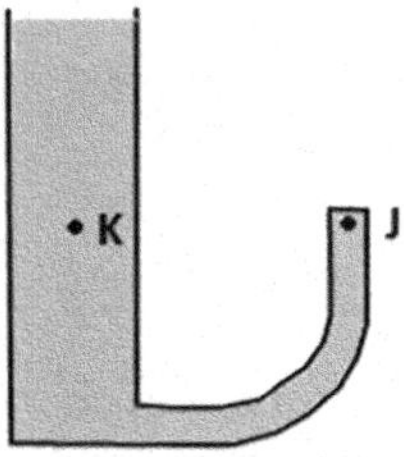

Q4. A nonviscous liquid of density $\rho = 1200$ kg·m^{-3} was flowing in a conduit between two circular sections A and B of area $A_B = 0.5\ A_A$. Section B is located at the same level of section A. In A the fluid pressure is $P_A = 60{,}000$ Pa and its speed is $v_A = 0.7$ m/s. What is the pressure of the liquid in B?

(A) 59,118 Pa (B) 60,221 Pa (C) 60,294 Pa

Q5. In a pipeline is flowing a fluid of density ρ. In a section A of the pipeline that is on the level h_A, the pressure is P_A and the speed is v_A. In a section B of the duct which is located at a level $h_A{-}h$, the pressure is $P_B = P_A$. What can be said of the relationship between the areas of the section A and section B?

(A) This ratio depends on h/v_A^2.

(B) This ratio depends on h, but does not depend on v_A.

(C) This ratio is independent both from v_A and from h.

Q6. In an open tube of rectangular cross section, there is a steady flow of water of thickness h. The velocity profile at different depths is shown in the figure. At what depth the water speed is half the speed with which it moves the surface layer of the water?

(A) h/2; (B) h; (C) h/4

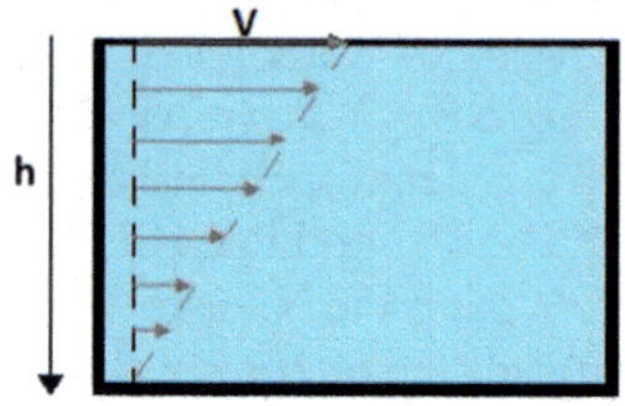

Q7. A steady flow of water flowing in a conduit. A U-tube, which contains mercury, is inserted between the sections A1 and A2 of the conduit, the diameter of which is a double of the other. What figure best represents the height of the mercury in the two branches of the U-tube?

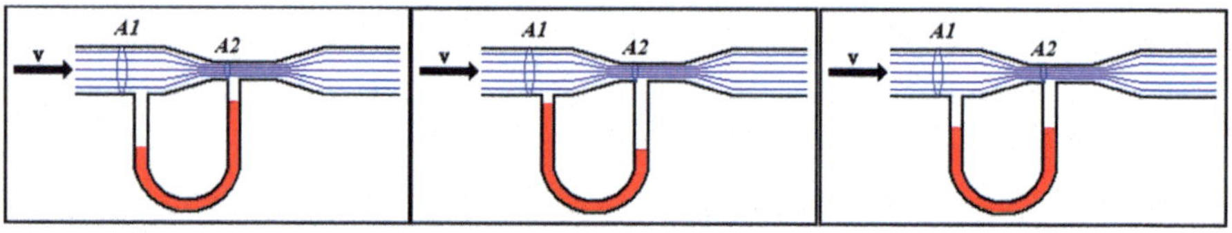

Q8. A glass ball is leaved on the surface of the water contained in a long vertical cylinder. Between the graphics shown on the right, which one best describes the time evolution of the speed of the ball when dropped into water?

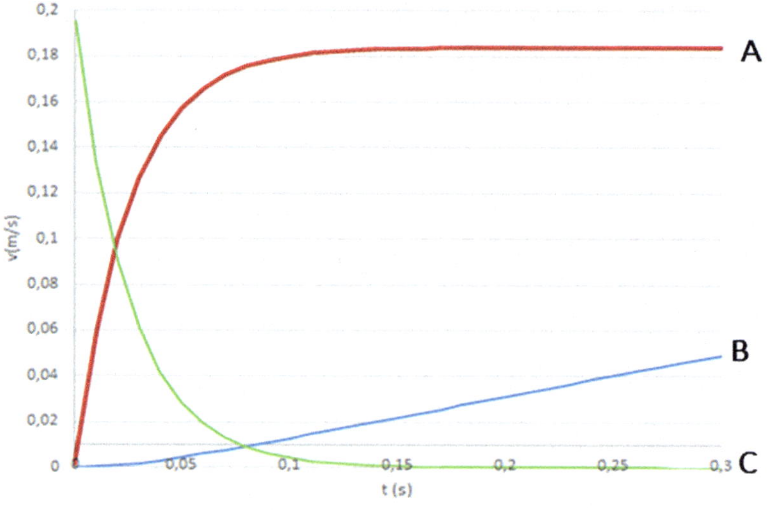

References

AAAS-American Academy for the Advancement of Science (2004). Invention and Impact: Building Excellence in Undergraduate Science, Technology, Engineering and Mathematics Education, Washington DC: AAAS.

AAAS-American Academy for the Advancement of Science (2011). Vision and Change in Undergraduate Biology Education, Washington, DC: AAAS.

Boix Mansilla, V, Duraisingh, E. D. (2007). Targeted assessment of students' interdisciplinary work: an empirically grounded framework. The Journal of Higher Education, 78 (2) 215–237.

Brewe, E., Pelaez, N. J., & Cooke, T. J. (2013). From Vision to Change: Educational Initiatives and Research at the Intersection of Physics and Biology. CBE—Life Sciences Education, 12, 117–119.

Bustamante, C. (2004). Of torques, forces, and protein machines. Protein Science, 13, 3061–3065.

CBE (2013). CBE—Life Sciences Education (LSE): describing teaching and learning at the intersection of biology and physics. CBE—Life Sciences Education, 12.

Collins, A, Joseph, D, & Bielaczyc, K. (2004). Design research: theoretical and methodological issues. J. Learn. Sci., 13, 15–42.

Cooper, M. M., & Klymkowsky, M. W. (2013). The Trouble with Chemical Energy: Why Understanding Bond Energies Requires an Interdisciplinary Systems Approach. CBE-Life Sci Educ., 12, 306–312.

Cummings, K., Laws, P.W., Redish E.F., Cooney, P.J., & Taylor, E. F. (2004). Understanding physics. Hoboken, NJ: Weley.

Donovan, D. A., Atkins, L. J., Salter, L. Y., Gallagher, D. J., Kratz R. F., Rousseauu, J. V., & Nelson, G. D. (2013). Advantages and Challenges of Using Physics Curricula as a Model for reforming an Undergraduate Biology Course. CBE-Life Sciences Education, 12, 215–229.

Dreyfus, B. W., Gouvea, J., Geller, B. D., Sawtelle, V., Turpen, C., & Redish, E. F. (2014). Chemical energy in an introductory physics course for life science students. American Journal of Physics, 82(5), 403–411.

Hall, K. L., Watkins, J. E., Coffey, J. E., Cooke, T.J., & Redish, E.F. (2011). Examining the impact of student expectations on undergraduate biology education reform. Paper presented at the American Educational Research Association National Meeting, held April 2011 in New Orleans, LA.

Heron, P.R. L., Shaffer, P. S., & McDermott, L. C. (2004). Research as a Guide to Improving Student Learning: An Example from Introductory Physics. In Invention and Impact: Building Excellence in Undergraduate Science, Technology, Engineering and Mathematics Education, Washington AAAS, pp. 33–38

Hoskinson, A. M., Caballero, M. D., & Knight, J. K. (2013). How Can We Improve Problem Solving in Undergraduate Biology? Applying Lessons from 30 Years of Physics Education Research. CBE— Life Sciences Education, 12, 153–161.

Hoskinson, A.M., Couch, B.A., Zwickl, B. M., Hinko, K., Caballero M.D. (2014). Bridging Physics and Biology Teaching through Modeling. American Journal of Physics, 82(5), 434–441.

Ivanitskaya, L., Clark, D., Montgomery, G., Primeau, R. (2002). Interdisciplinary learning: Process and outcomes. Innovative Higher Education, 27 (2) 95–111.

Laws, P. W. (2004). Promoting the Diffusion of Undergraduate Science Curriculum Reform: The Activity-Based Physics Suite as an Example. In Invention and Impact: Building Excellence in Undergraduate Science, Technology, Engineering and Mathematics Education, Washington AAAS, pp. 247–252. https://www.aaas.org/sites/default/files/08_Cre_App_Laws.pdf

Loverude, M. E., Heron, P. R. L., and C. H. Kautz (2010). Identifying and addressing student difficulties with hydrostatic pressure, American Journal of Physics 78, 75–85.

Manthey, S., & Brewe, E. (2013). Toward University Modeling Instruction—Biology: Adapting Curricular Frameworks from Physics to Biology, CBE-Life Sciences Education, 12, 206–214.

McDermott, L. C., Shaffer, P. S. (1992). Research as a guide for curriculum development: An example from introductory electricity. Part I: Investigation of student understanding. American Journal of Physics, 60 (11), 994–1003.

McDermott, L. C., Heron, P. R. L., Shaffer, P. S., & Stetzer M. R. (2006). Improving the preparation of K-12 teachers through physics education research, Am. J. Phys. 74 (9) 763–767.

Meredith, D.C., Redish, E.F. (2013). Reinventing physics for life-science majors. Physics Today, 66 (7) 38–43.

Michelini, M., & Stefanel, A. (2016) Teaching/Learning physics to non-physicist: physics for Agricultural, Biotech and Environmental. Contribution to GIREP Congress 2015, Wochlaw, July 6–10, 2015, in Proceedings of the GIREP Congress 2015.

Nersessian, N. J. (2002). The cognitive basis of model-based reasoning. In P Carruthers, S Stich, & M. Siegal (Eds.) The Cognitive Basis of Science. Cambridge, UK: Cambridge University Press, 133–153.

O'Shea, B., Terry, L., & Benenson, W. (2013). From F=ma to Flying Squirrels: Curricular Change in an Introductory Physics Course. CBE-Life Science Education, 12, 230–238.

Redish, E. F., Cooke, T. J. (2013). Learning each other's ropes: negotiating interdisciplinary authenticity. CBE-Life Sciences Education, 12, 175–186.

Redish, E. F., Hammer, D. (2009). Reinventing college physics for biologists: explicating an epistemological curriculum. Am. J. Phys. 77, 629–642.

Redish, E.F., Bauer, C., Carleton, K.L., Cooke, T.J., Cooper, M.M., Crouch, C.H., Dreyfus, B.W., Geller, B., Giannini, J., Svoboda Gouvea, J., Klymkowsky, M.W., Losert, W., Moore, K., Presson, J., Sawtelle, V., Thompson, K. V., Turpen, C., Zia, R.K.P. (2014). NEXUS/Physics: An interdisciplinary repurposing of physics for biologists. Am. J. Phys. 82, (5) 368–377.

Svoboda Gouvea, J., Sawtelle, V., Geller, B. D., and Turpen, C. (2013). A Framework for Analyzing Interdisciplinary Tasks: Implications for Student Learning and Curricular Design. CBE—Life Sciences Education, 12, 187–206.

Thompson, K. V., Chmielewski, J., Gaines, M. S., Hrycyna, C.A., & LaCourse, W. R. (2013). Competency-Based Reforms of the Undergraduate Biology Curriculum: Integrating the Physical and Biological Sciences. CBE—Life Sciences Education, 12, 162–169.

Watkins, J., & Elby, A. (2013). Context Dependence of Students' Views about the Role of Equations in Understanding Biology. CBE-Life Sciences Education, 12, 274–286.

Watkins, J., Coffey, J. E., Redish, E.F., Cooke, T.J. (2012). Disciplinary authenticity: enriching the reforms of introductory physics courses for life-science students. Phys. Rev. ST, Phys. Educ. Res., 8, 010112.

Chapter 11
The Design of Activities Based on Cognitive Scaffolding to Teach Physics

Genaro Zavala

Abstract Physics education research has produced many educational strategies such as Peer Instruction, Tutorials in Introductory Physics, and Real-Time Physics which one of the main objectives is to help students organize a better knowledge structure. For some settings, because equipment is needed for some of those strategies, then they are difficult to implement. This contribution recommends designing cognitive scaffolding activities without equipment in order to help students to choose the right scientific concept for the problem at hand. Cognitive scaffolding activities include strategies (i.e., by pumping questions) that help students reflect, think, or conceptualize. I present the framework in what the activities are based and describe a process of the construction of an activity for the understanding of the superposition principle applied to electric fields in a typical electricity and magnetism, first-year university course. At the end I present results implementing the activity with students by showing students' responses and reasoning to a problem in a midterm exam.

11.1 Introduction

In university-level teaching of physics, there are a number of active learning strategies that are based either on a format with a large number of students such as Peer Instruction (Mazur 1997) or on a format with a small number of students like laboratory-based strategies such as Tutorials in Introductory Physics (McDermott et al. 1998) and Real-Time Physics (Sokoloff et al. 2004). In many Latin American universities, the setting in which physics is taught is neither of those. In many of these universities, the number of students in sections is between 30 and 40. Moreover, classrooms in which these courses are taught could be in different buildings through the university. Although Peer Instruction or even Tutorials in Introductory

G. Zavala (✉)
School of Engineering and Sciences, Tecnologico de Monterrey, Monterrey, Mexico

School of Engineering, Universidad Andres Bello, Santiago, Chile
e-mail: genaro.zavala@itesm.mx

M. Pietrocola (ed.), *Upgrading Physics Education to Meet the Needs of Society*,
https://doi.org/10.1007/978-3-319-96163-7_11

Physics have been proven to be effective in this kind of settings, sometimes it is unpractical to have equipment needed for the strategy, clickers in the case of Peer Instruction and some equipment or a lot of that in the case of Tutorials in Introductory Physics and Real-Time Physics, respectively. We have been working on the design of problems based on cognitive scaffolding to teach physics. These problems are designed to be used in almost any setting since no equipment is needed. Students work in collaborative groups of three or four students each. The design consists of either (1) taking alternative conceptions students normally have and design a tutorial-format conceptual sequence to organize the knowledge structure of students related to that conception and (2) transforming a traditional problem to a tutorial-format problem which takes the student through scientific reasoning steps to build concepts, that is, a type of cognitive scaffolding (Mackiewicz and Thompson 2014). The objective of this contribution is to show the framework in what the activities are based and, at the same time, describe a process of the construction of an activity.

11.2 Background

In order to solve problems, students should choose from their own knowledge structure what concepts are related to the problem at hand and then choose a learned solving path to get to the solution. However, many times the students' knowledge structure is ill-defined with both scientific and alternative conceptions coexisting at the same time which make difficult to students to choose the right conception needed for the problem (Posner et al. 1982). The alternative conceptions students might acquire in their daily life or learned from not well-understood education during their previous years at school (Hammer 1996). Some research suggests that if a student chooses the scientific conception for the problem, there is more probability that the right solution is obtained, but if he/she chooses an alternative conception, then the result might not be the correct (Ling and Singh 2015).

There are many educational strategies such as Peer Instruction (Mazur 1997), Tutorials in Introductory Physics (McDermott et al. 1998), and Real-Time Physics (Sokoloff et al. 2004), to name a few, which one of the main objectives is to help students organize a better knowledge structure by working in pairs or in collaborative groups. For some settings, those strategies are difficult to implement. There is then a need to design activities in which instructors can implement without equipment. One way to design activities in order to help students to choose the right scientific concept for the problem at hand is using what has been called cognitive scaffolding (Mackiewicz and Thompson 2014) or simply scaffolding (Wood et al. 1976). Paraphrasing Wood, Bruner, and Ross, scaffolding is any process that enables the student to achieve an understanding of a scientific conception, solve a problem, carry out a task, or achieve a goal which would be beyond him/her if not assisted. Scaffolding could be given by the instructor or by peers, but also it could be given in the activity they solve. Cognitive scaffolding includes strategies (i.e., by pumping questions) that help students reflect, think, or conceptualize (Mackiewicz and Thompson 2014). In physics education research, scaffolding or cognitive scaffolding has been used for some years (Cui et al. 2005; Singh 2008; Garza and Zavala 2010;

Lisdstrom and Sharma 2011; Ding et al. 2011; Roll et al. 2012; Leinonen et al. 2013; Ling and Singh 2015).

11.3 Design of Activities

We have worked in two types of activities in which we used cognitive scaffolding. The first type of activities helps students to acquire scientific conceptions instead of alternative conceptions they might have. The main objective for this type of activities is precisely substitute, if possible, a strong alternative conception by a scientific-accepted conception. For this type of activities, we follow the following chart (Fig. 11.1).

In brief, (1) an alternative conception is chosen from the vast literature in the topic or from the experience that the instructor has with their own students during the years. The conception should be something that it is well known and strong enough that students could hold into it even after instruction. (2) A literature review is always a good practice to do since research articles not only normally explain what the alternative conception is related to and some actions one could take but also, from these articles, a situation could be found or ideas to think about a physical

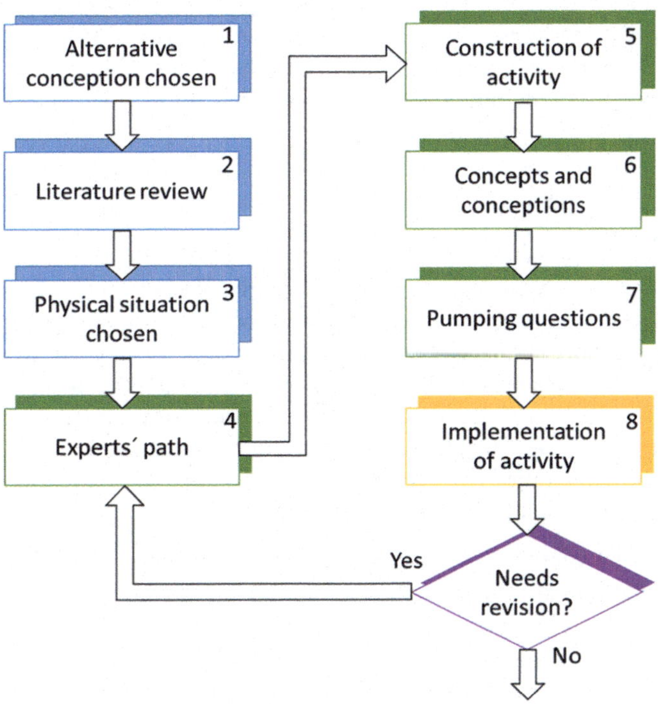

Fig. 11.1 Sketch of how an activity is designed to acquire an accepted scientific conception. The iteration could be as many as needed to have an activity that succeeds

situation could be found. Then, (3) a physical situation is chosen according to the alternative conception. It is important that the situation is familiar to students either from their daily life or academically. (4) The following step is to think and design an expert's path to solve the situation. In this case, the path could be made of a few questions or more, depending on what initially students understand. With the expert's path, (5) construct the activity trying to follow that path as close as possible. In each question, think of additional (6) concepts and conceptions students should have to continue. If students have acquired those concepts, continue. If not, then you have to propose one or more questions in which this concept or conception is acquired. (7) Design some pumping questions to make students reflect on their own learning before continuing. This could be achieved by questions that make them analyze their own answers and/or contrast two answers to different parts of the activity. (8) The last step before assessing is implementing the activity in a real situation inside the classroom. A better way to have feedback on how the activity helps students to acquire a scientific conception is to implement and interact with students to get that feedback. After implementing, you should ask yourself whether the activity needs revision or not. If it does, then return to review the expert's path and go on from there. If the activity was a success, then it might be ready.

The second type of activities takes a traditional end-of-the-chapter problem and transforms it into a problem in which scaffolding is given. The design process is similar to the previous type. However, in this case, the situation is what the problem presents. Another difference is that in these types of activities, the objective is not a specific alternative conception, but to help students to refine their problem-solving skills by understanding the scientific conceptions they should choose in their process.

11.4 Examples of Activities

The example in this section is an activity in the topic of electrostatics of the first type presented above. The activity tries to help students to understand the scientific conception of a conducting neutral object in terms of its capacity to produce an electric field. Some students have the alternative conception that neutral conducting objects (or in general that neutral objects) are not able to produce electric field no matter what. Garza and Zavala (2010) worked on this conception trying to see whether scaffolding improved students' understanding. They presented evidence that, in effect, some scaffolding produced better results.

Contextual Background The activity can be implemented in a university introductory electricity and magnetism class. The first topic in the course is electrical interactions in electrostatics. The activity needs students to be familiar with the induction of charge and the condition for electrostatic equilibrium as well as the superposition principle. The activity is implemented with students in any setting but in collaborative groups of three students each. This ensures that not only the activity or the instructor with interactions with groups provides scaffolding but also that peers also provide some scaffolding (Ge and Land 2003).

Description of the Activity The activity is presented in Appendix (without blank spaces that normally need student's responses). The activity starts with Section I which has the objective to present the simplest case, an isolated neutral conducting object. Students, in general, do not have a problem with the correct answer in this section since, even those with the alternative conception will have this answer correct. However, there are three questions which serve as cognitive scaffolding since the instructor and peer interactions make students understand that the cube is by itself, i.e., far away from any other object; that even though the object is neutral, it has charge (the same number of positive and negative charges); and that this charge is free to move if an electric force is present. The second question then is introduced and is expected that students answer that there is no electric field at location P and this answer is reserved for contrasting later on the activity.

Section II presents another situation with a nonconducting object with excess charge on the surfaces. The first questions are similar to those in Section I to make them reflect on what a nonconducting object is and that the sum of the negative and positive charge is zero. The situation serves as scaffolding since later on, when there is an induced charge on a conducting object, the distribution of charge will be similar to the object in this section. Students must get to the conclusion that, even that the object is neutral, because the negative charge is closer to the location, then, the electric field will exist at that location.

Section III is the heart of the activity. First, a point charge is isolated in the space, and because students have previously seen the electric field produced by point charges, then the question of the E-field at location S is not a main issue. Then, a neutral conducting object is placed next to the point charge. Question 2 is an example of resorting to concepts and conceptions that are not the main subject of the activity but that we think it is important to reinforce. In the same way is question 3 since it serves to resort to a concept that students should have and reinforce that concept as in question 2 in this section. However, it is also a pumping question for scaffolding since the charge distribution of students should be similar to the charge distribution presented in section II. Question 4 is a pumping question that serves as part of cognitive scaffolding since it makes students reflect on their own thinking and whether that thinking is consistent to previous questions. Question 5, similar to questions 2 and 3, is another question to reinforce concepts and conceptions. Question 6 is another example of scaffolding since it helps students to think about the electric field produced by the charge but also the electric field produced by the cube. Lastly, question 7 is another example of cognitive scaffolding since students reflect on a main issue of the activity, that is, the neutral object producing a field.

Section IV is the conclusion part of the activity. Questions 1 and 2 have different answers of the electric field produced by the same object which is neutral and conducting. Therefore, students should realize that, in the second case, there is an electric field produced by the neutral object. The last question is cognitive scaffolding since it makes students reflect on their own answers and come to a correct conclusion.

In a broader view of scaffolding, that is, construction of concepts and conceptions from the simpler ones to more sophisticated ones, the following activity that an instructor might assign is similar to the previous described activity, but in this other case, instead of having a neutral conductor, the new activity might use a neutral nonconducting object. At the end of the new activity, students understand that in the case of nonconducting materials, the polarization plays a similar role to the induction of charges in conductors.

11.5 Students' Responses in a Summative Evaluation

The activity presented is successful for students to acquire the scientific conception of electric fields produced by neutral conducting objects. In a midterm evaluation a month later, a similar problem was presented for students (see Fig. 11.2).

There were a total of 31 students who were in the course and presented the exam. Out of 31 students, 10 of them answered the 3 questions correctly, and 12 of them

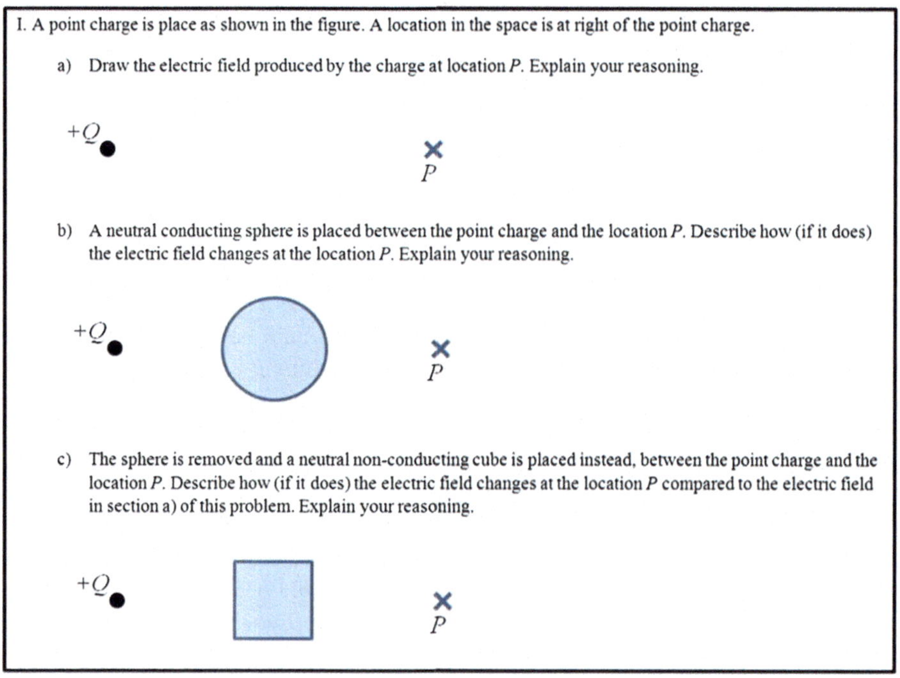

Fig. 11.2 Example of a problem in a midterm exam in an electricity and magnetism course. The problem asks for the electric field when a neutral conductor and a neutral nonconducting material are placed near a charge

answered the first 2 questions correctly but failed to correctly answer the third question.

Some students were able to answer the questions in a very elaborate way showing a complete understanding of the situation. An example of the answer of a student to question (b) is the following:

> The field increases since when we calculate the field by superposition we have the same charge at the same distance. However, there are additional induced charges in the sphere. These charges are identical in amount but positive on the right surface and negative on the left surface of the sphere. However, the positive charge is closer so its electric field at P is greater than the one produced by the negative charge, leaving a net electric field produced by the induced charge as it were a positive charge. The electric field at P would be as E produced by the point charge and E produced by the sphere which the sum is greater than before.

This student drew positive and negative charges on the right and left part of the sphere, respectively along with his answer. His answer is elaborate describing the process of induction of charge in the sphere and how that charge produces an additional source of electric field in the space. He even based his answer explaining explicitly that he uses the superposition principle.

Another less elaborate answer of another student is the following:

> The field increases. The charge polarizes the conducting sphere such that there will be positive charges on the right of the sphere and negative on the left that produce a field at P. By superposition the fields add together.

This student drew positive and negative charges on the right and left part of the sphere and an additional arrow to the right at the location *P* along with his answer.

All correct answers and reasoning for part (b) (22 out of 31) are written similar to an elaborate answer as the first student presented or less elaborate answers as the second student presented. There were seven students who answered correctly that the electric field increases but failed to explain, in a coherent way, why. The remaining two students answered that the field does not change. Their reasoning was similar saying that the electric field passes through the sphere without changing when "arriving" to the location *P*.

Part (c) of the problem (placing a nonconducting object) was harder for students. Ten students answered both questions correctly. One student wrote:

> The field will be less than when the conductor were place and greater than when nothing was there (part a). In this case a polarization is presented in the cube and it will produce a small field at location P adding to the one produced by the point charge in the same direction.

This student drew small dipoles in the cube representing the polarization. There were 12 students who coherently answered part (b) of the problem but failed part (c). From them there were some students who explained that in this case there was no induced charge since charges in nonconducting materials are not free to move. However, there were some other students that talked about polarization but even that, they mentioned that the field did not change.

11.6 Final Comments and Conclusions

We have presented the framework and design of cognitive scaffolding activities that help students to understand the scientific conceptions. The activities can be designed to acquire the scientific-accepted conception or to help students to choose the right conceptions in order to solve problems. They are scaffolding activities since they provide a support for students to achieve academic goals which otherwise would have been difficult to achieve. They are cognitive scaffolding since the activities make students reflect on their own understanding.

The implementation of the activities might be made as introduction to the topic or concept. Then later on that topic, students might work with activities with less scaffolding so that students are being prepared to let them alone during the homework assignments or exams. If an activity is related to a more sophisticated concept or problem-solving, then it can also be implemented not only at the beginning of the topic.

These activities help some students to acquire a robust understanding of scientific-accepted explanations of phenomena. An evidence of understanding was presented in which students answered the exam questions with elaborate reasoning.

A recommendation is to use these activities but monitor the way they are implemented since not all students have difficulties that the activity poses. Although we have some evidence that top students appreciate the activities since they serve them as an opportunity to better organize their own knowledge structure and because some of these students enjoy leading their teams, however, there is also evidence that sometimes they are not as effective for top students since the activity covers topics he/she has already learned (van Merrienboer and Sweller 2005).

Appendix

Section I

The figure shows an isolated neutral conductor cube.

1. Reflect on:

   (a) What an isolated means.
   (b) What it means that the cube is neutral.
   (c) What it means that the cube is a conductor. Elaborate on each answer.

Next to the cube, a location *P* in the space is indicated.

2. Is there an electric field in the location *P*? If so, sketch the electric field and explain your answer. If not, state so explicitly and explain why.

Section II

The figure shows an isolated neutral nonconducting cube with excess positive charge on the left and excess negative charge on the right.

1. Reflect on:
   (a) What it means that the cube is neutral and how that is related to the amount of charge on each side.
   (b) What it means that the cube is a nonconducting. Elaborate on each answer.

   Next to the cube, a location R in the space is indicated.

2. Is there an electric field in the location R? If so, sketch the electric field and explain your answer. If not, state so explicitly and explain why.

Section III

The figure shows an isolated point positive charge. Next to the charge, a location S in the space is indicated.

1. Sketch the electric field at S and explain your answer.

   A neutral conductor cube is placed next to the point charge as seen in the figure.

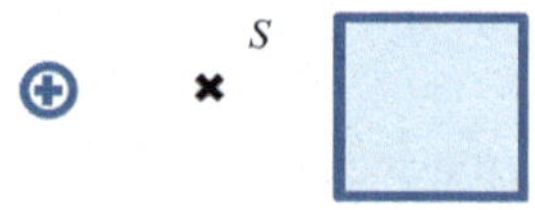

2. Electrostatic equilibrium is reached in an instance. What is or are the conditions for reaching the electrostatic equilibrium? Please elaborate.
3. Draw the distribution of the induced charge in the cube. Explain your reasoning.
4. Reflect on whether the cube continues to be neutral or not. If not, explain why. If it does it, see whether that answer is consistent to your charge distribution drawing in question 3 of this section.
5. State the superposition principle for the electric field.
6. Taking into account the superposition principle, answer whether the electric field at location S in question 1 of this section is modified or not compared to the electric field at location S when the neutral conductor was placed. If it is modified, explain how. If it is not, explain why not.
7. Reflect on the electric field produced at S exclusively by the neutral conductor. Describe that electric field at location S.

Section IV

Take your answers to question 2 of Section I and question 7 of Section III.

1. In question 2 of Section I, there was an isolated neutral conductor cube; does this neutral conductor produce an electric field?
2. In question 7 of Section III, there was a neutral conductor cube; does this neutral conductor produce an electric field?
3. Comparing the answers to questions 1 and 2 in this section, how would you respond to the question: is it possible that a neutral conducting object produces an electric field? Explain your reasoning.

References

Cui, L., Rebello, N.S. & Bennett, A.G. (2005). College students' transfer from calculus to physics. *AIP Conference Proceedings, 818,* 37–40.

Ding, L., Reay, N., Lee, A. & Bao, L. (2011). Exploring the role of conceptual scaffolding in solving synthesis problems. *Physical Review Special Topics-Physics Education Research, 7,* 020109.

Garza, A. & Zavala, G. (2010). Electric Field Concept: Effect of the Context and the Type of Questions. *AIP Conference Proceedings, 1289*, 145–148.

Ge, X. & Land, S. M. (2003). Scaffolding students' problem solving processes in an ill-structured task using question prompts and peer interactions. *Educational Technology Research and Development, 51*, 21.

Hammer, D. (1996). More than misconceptions: Multiple perspectives on student knowledge and reasoning, and an appropriate role for education research. *American Journal of Physics, 64* (10), 1316–1325.

Leinonen, R., Asikainen, M. A. & Hirvonen, P. E. (2013). Overcoming students' misconceptions concerning thermal physics with the aid of hints and peer interaction during a lecture course. *Physical Review Special Topics-Physics Education Research, 9,* 020112.

Lin, S-Y & Singh, C. (2015). Effect of scaffolding on helping introductory physics students solve quantitative problems involving strong alternative conceptions. *Physical Review Special Topics-Physics Education Research, 11,* 020105.

Lisdstrom, C. & Sharma, M. D. (2011). Teaching physics novices at university: A case for stronger scaffolding. *Physical Review Special Topics-Physics Education Research, 7,* 010109.

Mackiewicz, J. & Thompson, I. (2014). Instruction, Cognitive Scaffolding, and Motivational Scaffolding in Writing Center Tutoring. *Composition Studies, 42* (1), 54–78.

Mazur, E. (1997). *Peer Instruction: A User's Manual.* Upper Saddle River, NJ: Prentice Hall.

McDermott, L. C., Shaffer, P. S. and the PEG (1998). *Tutorials in Introductory Physics.* Englewood Cliffs, NJ: Prentice-Hall.

Posner, G., Strike, K, Hewson, P. & Gerzog, W. (1982). Accommodation of a scientific conception: Toward a theory of conceptual change. *Science Education, 66*, 211.

Roll, I., Holmes, N. G., Day, J. & Bonn, D. (2012). Evaluating metacognitive scaffolding in Guided Invention Activities. *Instructional Science, 40* (4), 691–710.

Singh, C. (2008). Assessing student expertise in introductory physics with isomorphic problems. II. Effect of some potential factors on problem solving and transfer. *Physical Review Special Topics-Physics Education Research, 4,* 010105.

Sokoloff, D. R., Thornton, R. K. & Laws, P. (2004). *RealTime Physics, Active learning laboratories.* New York: John Wiley & Sons.

van Merrienboer, J. J. G. & Sweller, J. (2005). Cognitive load theory and complex learning: Recent developments and future directions. *Educational Psychology Review, 17*, 147.

Wood, D., Bruner, J. S. & Ross, G. (1976). The Role of Tutoring in Problem Solving. *Journal of Child Psychology and Psychiatry, 17* (2), 89–100.

Chapter 12
Examining the Relationships Among Intuition, Reasoning, and Conceptual Understanding in Physics

Mila Kryjevskaia

Abstract It is a common expectation that, after instruction, students will consciously and systematically construct chains of reasoning that start from established scientific principles and lead to well-justified predictions. When student performance on course exams does not reveal such patterns, it is often assumed that students either do not possess a suitable understanding of the relevant physics or are unable to construct such inferential reasoning chains due to deficiencies in reasoning abilities. Psychological research on thinking and reasoning, however, seems to suggest that, in many cases, thinking processes follow paths that are strikingly different from those outlined above. A set of theoretical ideas, referred to broadly as dual process theory, asserts that human cognition relies on two largely independent thinking systems. The first of these systems is fast and intuitive, while the second is slow, logically deliberate, and effortful. In an ongoing project focusing on student reasoning in physics, we have been developing and applying various methodologies that allow us to disentangle reasoning, intuition, and conceptual understanding in physics. We then use the dual process theory to account for the observed patterns in student responses. Data from introductory physics courses are presented and implications for instruction are discussed.

12.1 Introduction

It is a common expectation that, after instruction, students will consciously and systematically construct chains of reasoning that start from established scientific principles and lead to well-justified predictions. When student performance on course exams does not reveal such patterns, it is often assumed that students either do not possess a suitable understanding of the relevant physics or are unable to construct such inferential reasoning chains due to deficiencies in reasoning abilities. Psychological research on thinking and reasoning, however, seems to suggest that,

M. Kryjevskaia (✉)
North Dakota State University, Fargo, ND, USA
e-mail: mila.kryjevskaia@ndsu.edu

M. Pietrocola (ed.), *Upgrading Physics Education to Meet the Needs of Society*,
https://doi.org/10.1007/978-3-319-96163-7_12

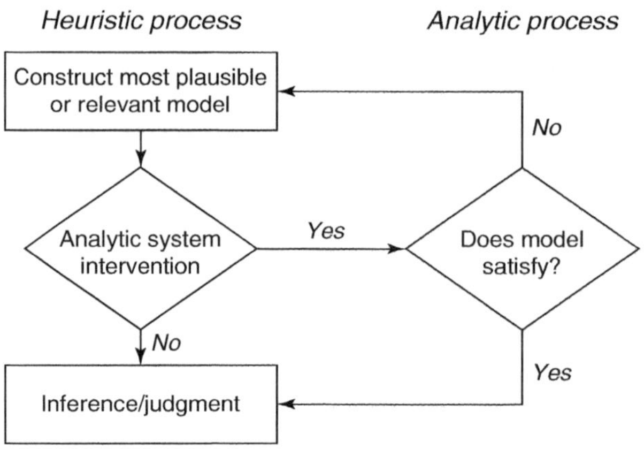

Fig. 12.1 Illustration of interactions between the heuristic and analytic processes (Evans 2006).

in many cases, thinking processes follow paths that are strikingly different from those outlined above. A set of theoretical ideas, referred to broadly as the dual process theory, asserts that human cognition relies on two largely independent thinking processes. The first of these processes is fast and intuitive (often referred to as the *heuristic* process), while the second is slow, logically deliberate, and effortful (often referred to as the *analytic* process). In an ongoing project focusing on student reasoning in physics, we have been developing and applying various methodologies that allow us to disentangle reasoning, intuition, and conceptual understanding. We then use the dual process theory to account for the observed patterns in student responses (Fig. 12.1).

The Dual Process Theory of reasoning suggests that two processes are involved in most cognitive tasks: heuristic and analytic. When a reasoner is presented with an unfamiliar situation, the quick and intuitive heuristic process immediately and subconsciously suggests a most plausible and relevant mental model for this situation. This "first available mental model" is based on the person's prior knowledge, experiences, and contextual cues. In everyday life, first available mental models are often described as "gut feelings" or "first impressions." In this study, we use the term intuition to refer to student ideas consistent with first available mental models suggested by the quick, subconscious, and automatic heuristic process. However, it is important to note that, in the context of physics instruction, first available mental models may not necessarily be based on students' everyday experiences. Such models may be based on formal ideas or reasoning approaches ubiquitous in physics, but not necessarily applicable to a situation at hand.

The role of the analytic process is to assess the validity of a first available mental model (See Fig. 12.1). However, if a reasoner feels confident in the answer suggested by the model, the analytic process is often bypassed. In such cases, this process yields a final, heuristic-based response. The engagement on the analytic process, however, does not always result in a rigorous and systematic assessment of the validity of a mental model due to reasoning biases. For example, if a reasoner believes that an answer suggested by a heuristic-based model is correct, the reasoner will search for evidence that supports what is already believed to be correct while neglecting to consider alternatives. This thinking pattern is often referred to as *confirmation bias*. While it has been argued that metacognition, or thinking about one's own thinking, is the key for engaging the analytic process more productively (Amsel et al. 2008), the mechanism for a productive evaluation of heuristic-based mental models is poorly understood. In this study we aim to probe conditions under which students are more likely to recognize inadequacies in their current mental models and to consider alternative solutions.

12.2 Methodology

In order to achieve our goal, it is imperative to design methodologies that would allow for the disentanglement of student conceptual understanding from their reasoning approaches. In the past several years, we have been designing sequences of screening and target questions in various physics contexts (Kryjevskaia and Stetzer 2014). Screening questions probe whether or not a student possesses the formal knowledge and skills necessary to analyze a specific situation correctly. A target question requires the application of the same knowledge and skills in a similar situation but may also elicit intuitive, rather than formal reasoning approaches. We then focus on analyzing responses to target questions of those students who answer the screening questions correctly, thereby demonstrating that they indeed possess the formal knowledge and skills necessary to successfully arrive at a correct answer to the target question. This approach allows us to insure that patterns of student responses on the target question are likely to be attributed to specific reasoning approaches rather than the lack of conceptual understanding.

Below we present a screening-target sequence of questions in the context of static friction administered in the first semester of introductory calculus-based physics course. We then discuss three different modes of metacognitive intervention along with the theoretical underpinnings that informed the design of these interventions. Results are interpreted through the lens of the dual process theory of reasoning.

12.3 Examples and Results

Original Screening-Target Sequence of Questions

On the screening question, students considered box A at rest on a rough surface. They were told that a horizontal 30 N force was applied to the box, as shown in Fig. 12.2a, and the box was observed to remain at rest. Students were asked to compare the magnitudes of the applied force and the force of friction. In order to answer the question correctly, students were expected to apply Newton's second law

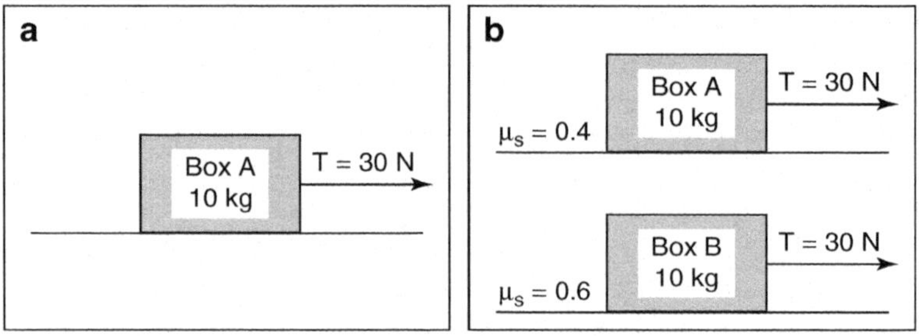

Fig. 12.2 Diagrams illustrating situations presented on (**a**) screening and (**b**) target questions

and to recognize that, since the box remains at rest, the net force on the box must be zero. As such, the magnitude of the force of friction must be equal to that of the applied force, $f_s = T = 30$ N.

The target question shown in Fig. 12.2b involved an analogous situation: two boxes of equal mass were placed on rough surfaces. Students were told that the coefficient of static friction μ_s between box A and a surface was 0.4, while the coefficient of static friction between box B and a different surface was 0.6. Identical 30 N horizontal forces were applied to each box; both boxes were observed to remain at rest. Students were asked to compare the magnitude of the force of friction acting on box A to that acting on box B. Both screening and target questions call for the application of the same reasoning approach: since both boxes remain at rest, the forces of friction must be equal to 30 N regardless of the roughness of the surfaces.

Most students were able to answer the screening question correctly, as shown in Table 12.1. However, ~23% of the students who applied the correct line of reasoning on the screening question failed to do so on the target question. Instead, these students argued that the force of friction on box A must be less than that on box B since $(\mu_s)_A < (\mu_s)_B$. Most of these students justified their answers by inappropriately applying various mathematical relationships between forces of friction and coefficients of friction such as $f_k = \mu_k N$ or $f_{s,\max} = \mu_s N$.

The application of the dual process theory suggests that the inclusion of the extraneous information on the screening question cued the mental model based on the relationships between the force of friction and the coefficient of friction. This resulted in the abandonment of the line of reasoning based on Newton's second law. The readily available and ubiquitous (in the context of introductory physics courses) mathematical relationships between these two quantities provided further confirmation for the validity of this mental model. As such, even though these students possessed the formal knowledge and skills necessary to answer the target question correctly, they did not feel compelled to examine the validity of their mental models either by checking for consistency between their answers to the screening and the target questions or by searching for alternative solutions.

The results from this screening-target sequence suggest that a fraction of incorrect student responses to the target question could be attributed to deficiencies in student reasoning approaches rather than the lack of knowledge and skills necessary to

Table 12.1 Results from student responses on sequences of screening-target questions

Original screening-target sequence ($N = 54$)		Metacognitive intervention 1 ($N = 53$)		Metacognitive intervention 2 ($N = 58$)		Metacognitive intervention 3 ($N = 224$)	
% of correct responses on the screening question							
81%		85%		57%		43%	
Distribution of responses on the target question of those students only who answered the screening question correctly							
Correct	Incorrect	Correct	Incorrect	Correct	Incorrect	Correct	Incorrect
77%	23%	76%	24%	76%	24%	73%	27%

answer the question correctly. As such, it is imperative to develop interventions that would engage students' analytic processes more productively and enable students to recognize shortcomings in their reasoning approaches. The three modes of metacognitive interventions described below were designed in order to probe impacts of various interventions on student reasoning patterns. All three interventions utilized the context of frictional force; the prevalence of incorrect student responses on the target question was used to gauge the effectiveness of these interventions.

Metacognitive Intervention 1: Opportunities for Considering Alternatives

In this sequence, a metacognitive question was designed to follow up the screening-target pair discussed above. The questions prompted students to (1) predict what answer other students would give if they applied intuitive thinking to the target question and (2) reflect on whether they themselves applied intuitive reasoning or formal knowledge. It is important to note that it was not the goal of this intervention to examine the students' abilities to distinguish between intuitive and formal thinking. Instead, we hoped that the metacognitive prompt would provide opportunities for the students to consider alternatives and to reflect on their own reasoning. This sequence was administered as part of a regular course exam. Students who responded to the metacognitive prompt received 1% of extra credit. The results from the metacognitive intervention 1, presented in Table 12.1, revealed no intervention-dependent difference in the student performance on the target question.

The results of metacognitive intervention 1 suggest that although this mode of intervention encouraged students to consider alternatives, it failed to create dissatisfaction with the students' current incorrect reasoning approaches. This apparent failure may be due to the state of *cognitive ease* that the screening-target sequence presents. Indeed, one of the functions of the heuristic process is to conduct a quick and subconscious assessment of a situation at hand. The heuristic process detects whether the situation presents any threat and whether or not some cognitive efforts must be redirected to the analytic process. If the heuristic process is not on alert (or is in the state of cognitive ease), the analytic process may not be engaged to its full potential. When the heuristic process detects "unease" (or is in the state of *cognitive strain*), the analytic process may be more fully engaged, such that a reasoner becomes more attentive and careful in his/her judgments (Kahneman 2011).

It may be argued that the screening question prompts students to apply the reasoning necessary to answer the target question correctly in a fairly simple context and, therefore, sets the students on the correct path in answering the target question. As such, the inclusion of the screening question makes it more likely for the students to answer the target question correctly. At the same time, it may be argued that the fairly straightforward solution to the screening question creates the state of cognitive ease. This, in turn, may impede student tendency to reflect on their thinking and to check for consistency between their reasoning on the screening and target questions. In other words, this state of cognitive ease may suppress student abilities to recognize reasoning pitfalls associated with the inappropriate application for the μ-based reasoning on the target question.

Metacognitive Interventions 2 and 3: Removing Cognitive Ease

Metacognitive interventions 2 and 3 were designed in order to probe whether or not the pattern of student responses on the target question could be altered by creating the cognitive strain on the screening question. Specifically, metacognitive intervention 2 contained a screening question that involved box A at rest on a rough surface with the coefficient of static friction $\mu = 0.5$. Students were told that a horizontal 30 N force was applied to the box, as shown in Fig. 12.3a, and the box remained at rest. The mass of box A was not specified. Students were asked to compare the magnitudes of the applied force and the force of friction. Students were prompted to state explicitly if not enough information was given to answer the question. The target question in this modified sequence was similar to that in the original version, except no masses for the two boxes were specified, as shown in Fig. 12.3b.

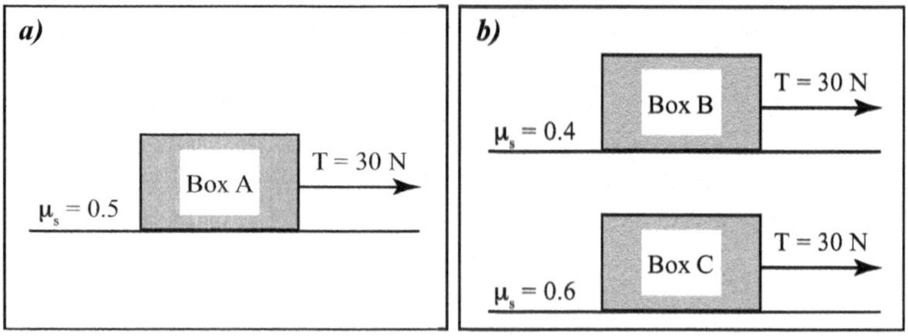

Fig. 12.3 Diagrams illustrating situations presented on (**a**) screening and (**b**) target questions of metacognitive intervention 2

We argued that this version of the screening question would create cognitive strain because the inclusion of the extraneous information (i.e., $\mu = 0.5$) would likely cue the reasoning approach based on the first available mental model "μ determines f," while the absence of information about the mass of the box would make it impossible to apply the mathematical relationship $f = \mu N$ inappropriately in order to justify the model and to determine the answer to the question. We hypothesized that the fraction of correct responses to this version of the screening question would decrease significantly, while the percentage of responses with inconsistent lines of reasoning on the screening and target questions would decrease. The latter hypothesis stemmed from the notion that those students, who correctly answer the screening question by rejecting the relevance of μ to the static situations, would not be likely to apply this rejected line of reasoning on the target question. Results presented in Table 12.1 suggest that our hypothesis was supported only partially. Indeed, while the percentage of correct responses on the screening question decreased, the pattern of inconsistent responses was not altered. On the target questions, ~20% of the students who answered the screening question correctly argued that "the higher coefficient of friction will lead to a great frictional force."

Metacognitive intervention 3 included a new version of the screening question designed to create cognitive strain by widening the space of possibilities through foregrounding the distinction between the cases of static and kinetic friction. Specifically, students considered a box at rest on a rough surface, which is observed to remain at rest after a hand exerted a horizontal force on the box, as shown in Fig. 12.4. Students were asked to identify which of the following piece or pieces of information are required in order to determine the magnitude of the force of static friction acting on the box: (a) the magnitude of the force exerted on the box by the hand, (b) the mass of the box, and (c) the coefficient of static friction between the box and the surface. Students were prompted to choose all that apply. The target question in this metacognitive intervention was identical to that in the original sequence. Much like on metacognitive intervention 2, we hypothesized that those students who correctly answer the screening question by rejecting the relevance of μ and the mass m to the static situations would not likely to apply this rejected line of reasoning on the target question. This version of the intervention, however, makes the rejection of the relevance of μ and m more explicit. Results presented in Table 12.1 suggest that all three modes of metacognitive intervention were unsuccessful in altering patterns of inconsistent student responses on the target question.

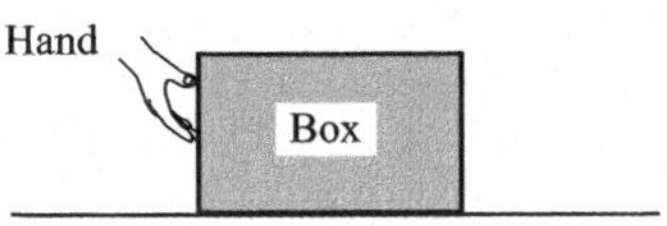

Fig. 12.4 A diagram illustrating a situation presented on a screening question of metacognitive intervention 3

12.4 Conclusions

The findings from our study serve to highlight the persistence and resilience of the kind of incorrect, intuitively appealing reasoning approaches often employed by introductory physics students. These observed intuitive approaches may not necessarily be based on everyday ideas related to a situation at hand (e.g., frictional force). In most cases even those students who demonstrated that they possessed the formal knowledge and skills necessary to answer the target question correctly did not feel compelled to examine the validity of their reasoning either by checking for consistency between their answers to the screening and the target questions or by searching for alternative solutions. Our results suggest that many students found confirmation of their intuitive ideas in misinterpreted formal mathematical relationships. As such, these students were particularly unlikely to question their first-impression answers.

We have presented data from three different metacognitive interventions designed on the basis of theoretical ideas rooted in the psychological research on reasoning and decision making and aimed at engaging students' analytic processes more productively. Despite these efforts, we have observed similar reasoning patterns on the target question in all three cases. This suggests that perhaps more targeted and systematic instructional approaches are needed to change students' habits of mind in the context of physics instruction.

References

D. Kahneman, *Thinking, Fast and Slow* (Farrar, Straus and Giroux, 2011).

J. St. B. T. Evans, *Psychonomic bulletin review*, **13**(3), 378–395 (2006).

E. Amsel, P A. Klaczynski, A. Johnston, S. Bench, J. Close, E. Sadler, and R. Walker, Cognitive Development **23**, 452–471 (2008).

M. Kryjevskaia, M. R. Stetzer, *Phys. Rev. Phys. Educ. Res.*, **10**, 020109, 2014.

Chapter 13
Conceptual Development and Critical Attitude in Physics Education: A Pathway in the Search for Coherence

Laurence Viennot

Abstract In recent years, the need to attract students towards physics has largely contributed to shape the objectives ascribed to physics teaching in many countries. Among these objectives, critical thinking is unanimously presented as of central importance. But at the same time a stress on competences jointly with this trend to simplify physics might lead to an underdevelopment of conceptual understanding. The following question then arises: can critical thinking be fostered in students without conceptual structuring and (still more importantly) without stressing the pivotal role of a search for coherence in science? This chapter briefly summarises some recent research on the joint development of a conceptual understanding and critical position among university students. In characterising students' responses when confronted to various explanations of a physical phenomenon, these studies bring to bear conceptual markers as well as meta-cognitive, affective and critical indicators. Some profiles of co-development are characterised, including "delayed critique" and "expert anaesthesia of judgment". The results strongly suggest that to disregard the objective of conceptual structuring is counterproductive for the development of students' critical attitude. Through these exploratory studies, it appears that the conditions in which students can begin to search for coherence—whether in pursuit of conceptual understanding or to activate their critical potential—constitute a crucial objective for further research.

13.1 Introduction

From the outset of physics education research in the 1970s, an implicit or explicit target was to facilitate better comprehension of accepted physics in students at all academic levels. But a consensus has long since been reached that what we call "accepted physics" had to be rethought for teaching, and Kattmann and Duit (1998) proposed the term "educational reconstruction" to designate this process. From this perspective,

L. Viennot (✉)
Laboratoire Matière et Systèmes Complexes, UMR 7057, Universitè Paris Diderot/CNRS, Paris, France
e-mail: laurence.viennot@univ-paris-diderot.fr

M. Pietrocola (ed.), *Upgrading Physics Education to Meet the Needs of Society*,
https://doi.org/10.1007/978-3-319-96163-7_13

the very basis of reflection about the design of a learning environment for a given audience combines examination of the content and of students' "prescientific conceptions". Of course, decisions about the design and implementation of learning environments are also influenced by the international context, with current consensus around a series of teaching goals: to engage students with physics and (and therefore) to simplify the contents to be taught in helping students to construct a preliminary sense of the nature of science, stressing the role of reasoning and reasoned inquiry and developing many other abilities—most prominently, critical stance. A stress on competencies can be observed in many official texts; for instance, the European Commission foregrounds the need "to develop the competencies for problem-solving and innovation, as well as analytical and critical thinking that are necessary to empower citizens to lead personally fulfilling, socially responsible and professionally-engaged lives" (European Commission 2015). Given this emphasis on competencies in combination with the desire to simplify physics, there is a risk that conceptual structuring may be disregarded, potentially resulting in serious inconsistencies in pedagogical resources. This in turn demands increased critical vigilance among students and teachers, inviting the research question *What links can be identified between the development of conceptual understanding and critical attitude in physics students*—or, in operational terms: *Can we help students to develop their critical thinking without a conceptual basis?*

13.2 Analysing the Interplay Between Conceptual Understanding and Critical Attitude: What Are We Talking About?

These questions identify a need to further specify two aspects of a student's intellectual trajectory—conceptual understanding and critical attitude—which are the focus of this chapter. As noted above, a process of *educational reconstruction* is needed to define what we mean by "conceptual understanding" of the content in play. The proposed framework (Viennot 2015a, 2016b; see Fig. 13.1) emphasises the importance of two factors. First, conceptual coherence is seen as pivotal in the process of educational reconstruction, echoing a vision of science as pursuing a coherent, predictive and parsimonious description of the material world (Jenkins 2007; Ogborn 1997). This search for coherence is explored in the content analysis, as well as in what we describe as *students' ideas*. Secondly, beyond students' existing ideas and ways of reasoning, it is proposed to take account of teaching rituals. As some of these rituals are found to be in a kind of resonance with students' ideas, the label "echo-explanation" is used to designate such cases—for instance, using "ray boxes" without any caveat (Viennot 2006). In any event, the persistence of these rituals over time (Viennot 2016) suggests that they are linked to factors that may influence the decisions of course designers and teachers. For that reason, these rituals and their likely determinants should be considered in any definition of a conceptual teaching target.

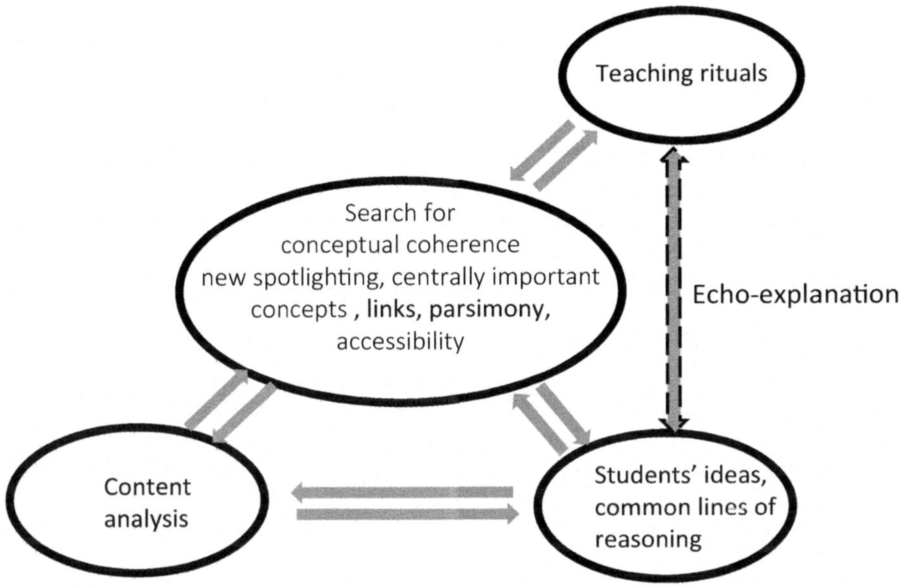

Fig. 13.1 A framework to define the conceptual content to be taught (Viennot 2015a, b)

On this view, the global coherence of a given conceptual target entails a particular "spotlighting" of the relevant content (Viennot 2003, 2015a). This means that explicit decisions must be made about centrally important concepts and—most importantly—about the links to be highlighted. The whole process should respect the epistemic need for parsimony and the constraint of accessibility.

The type of critical attitude considered here involves localising an explanation's possible incoherence or detecting its possible incompleteness. Socio-scientific forms of critique such as the ability to evaluate texts and their sources in terms of possible asymmetries of power or the status of experts (Jimenez and Puig 2012) are not examined here.

That said, an investigation of how students express or fail to express possible criticisms of an explanation does not exclude other psycho-cognitive factors. Clearly, posing a critical question requires some awareness of one's own state of comprehension and of what it is to learn science. Such questioning also evidences a search for intellectual satisfaction (often manifesting as frustration) (Viennot 2006; Mathé and Viennot 2009) and depends in part on such factors as self-esteem or self-efficacy (Bandura 2001). As these metacognitive and affective components of students' critical attitude seem a priori difficult to disentangle, they are designated here by the compound label "metacognitive-critical-affective" (MCA).

The investigations reported below examined how these MCA factors may evolve in conjunction with conceptual comprehension, focusing on students' *intellectual dynamics of co-development* during an interaction targeting conceptual progress. These studies entailed fine-grained analyses of *changes* in students' intellectual pathways over time.

13.3 A Few Investigations

Each of the reported studies centred on a particular physics topic, and several involved extended individual interviews (Mathé and Viennot 2009; Décamp and Viennot 2015; Viennot and Décamp 2016a, b). In each case, conditions were designed for a *concept-driven interactive pathway* (CDIP) (see Viennot and de Hosson 2015). A CDIP is a series of events—inputs from the interviewer and responses from the student, possibly involving experiments, questions or requests and discussion—informed by a search for coherence and intellectual satisfaction. In line with the teaching experiment method (Komorek and Duit 2004), the interaction is structured and guided to allow students to express their initial thoughts and reactions to various events. A CDIP is also progressive, in that what is understood at a given step may serve to construct the next stage of knowledge, offering opportunities for students to critique a few explanations. The three studies summarised here delineate a preliminary but consistent landscape constituted by the key intellectual dynamics of students' co-development.

Radiocarbon Dating

Although not the first in the series, a study centred on the topic of radiocarbon dating (Viennot 2014a; Décamp and Viennot 2015) provides a good illustration of our initial results. Beyond the well-known exponential decay of radiocarbon in dead organisms and the role of ^{14}C half-life (5730 years), a *relatively* complete and coherent explanation of this process should include at least the following conceptual nodes:

1. The need to know the initial proportion of radiocarbon to ordinary carbon in an organism at the time of its death
2. The uniformity of this quantity in the atmosphere and living beings
3. The constancy in time of this quantity
4. The process of formation of radiocarbon
5. The process of radioactive decay of radiocarbon
6. How the balance between the corresponding numbers per second of radiocarbon atoms involved in these two processes results in a steady value of $[^{14}C/^{12}C]$ in the atmosphere
7. The constancy of the total number of nuclei (radiocarbon + nitrogen)
8. The multiplicative effect of the existing numbers of radiocarbon and nitrogen nuclei in the destruction and creation of ^{14}C nuclei, respectively
9. How this multiplicative structure explains the stable proportion of radiocarbon to ordinary carbon in the atmosphere

We selected five online documents providing explanations that were incomplete with regard to the above list. Using conceptual nodes 7, 8 and 9, we also designed a sixth document to explain how a steady-state ^{14}C population can be reached and maintained from an unbalanced initial situation. Ten prospective teachers were then presented with these documents in order of increasing completeness. For each document (i.e. at each "step" from 1 to 6), students were asked to indicate their level of satisfaction with the explanation provided and whether they would need further information. Questions that might have been posed in response to the first and least complete explanations would include "How is it that there is a constant proportion of radiocarbon in the atmosphere? There is no radiocarbon decay in the atmosphere?"

Transcripts were processed at two levels of analysis: a conceptual level (which is not commented on here) and MCA aspects. Our MCA indicators included levels of agreement, types of questions posed (Table 13.1)—specifically, anecdotal or "crucial" (i.e. concerning the above conceptual nodes)—and levels of intellectual satisfaction or frustration (Table 13.2).

With the criteria we used to rank the students in Tables 13.1 and 13.2, the same in the two tables, and excluding the two last students (Y and H), we can observe that the same "diagonal" (broadly speaking, from col. 3, line 2, to col. 6, line 9) divides the tables in two parts.

Left of this diagonal, we find only questions about details and agreement or half-hearted agreement (Table 13.1) along with comments expressing satisfaction (Table 13.2) despite the incompleteness of the explanations:

> M, text 1: Yes, no, no, there are some good hints. It's true that it's very succinct, but you cannot give a whole course ... Err, no, no, it's clear, concise. (...) It's very complete.
>
> J, text 1: No, err, it's rather clear too. We understand the principle at least.

Table 13.1 Level of agreement after each step of the interview and questions posed

Interviewee	S_1	S_2	S_3	S_4	S_5	S_6	S_7
G	$\approx$ dl	$\approx$ **cq**	**cq**^{2}	**cq**	**cq**	Θ	3
B	Θ	Θ **cq**	**cq**^{3}	**cq**	**cq**^{3}	Θ	4
S	$\approx$ dl	$\approx$ dl^3	$\approx$	**cq**^{2}		Θ	3
J	Θ	Θ	Θ dl^2	**cq**^{2}	**cq**	Θ	3
M	Θ	Θ	Θ	dl **cq**	**cq**	Θ	3
T	$\approx$ dl	$\approx$ dl	Θ	**cq**	$\approx$	Θ	2
A	$\approx$	Θ	Θ		**cq**	Θ	2.5
V	Θ	Θ	Θ dl	$\approx$	**cq**^{2}	Θ	3
Y	$\approx$ dl	Θ dl	Θ	Θ	Θ	Θ	4
H	Θ	Θ dl	Θ dl^2	Θ	Θ	Θ	4

Abbreviations: Θ, $\approx$, agreement, half-hearted agreement; *dl* question about "details"; ***cq*** crucial question and shaded box; last column, final intellectual satisfaction, Likert scale from 1 (low) to 4 (high)

Table 13.2 Level of agreement at the end of each step and statements expressing satisfaction or frustration

Interviewee	S_1	S_2	S_3	S_4	S_5	S_6	S_7
G	≈	≈ **m**−	**m**−	**m**−	**m**−^{2}	Θ **m**+	3
B	Θ	Θ **m**+	**m**+ **m**−^{2}	**m**−	**m**−	Θ **m**+	4
S	≈	≈	≈	**m**−	**m**−	Θ **m**+	3
J	Θ	Θ **m**+	Θ	**m**−	**m**−	Θ **m**+	3
M	Θ	Θ **m**+	Θ	**m**−	**m**−	Θ **m**+	3
T	≈ **m**−	≈ **m**−	Θ		≈ **m**−	Θ **m**+	2
A	≈	Θ **m**+	Θ **m**+^{2}	**m**−	**m**−	Θ **m**+	2.5
V	Θ	Θ	Θ	≈	**m**−	Θ	3
Y	≈	Θ **m**+	Θ **m**+	Θ	Θ	Θ **m**+	4
H	Θ	Θ	Θ	Θ	Θ	Θ	4

Abbreviations: *Θ*, ≈, agreement, half-hearted agreement; **m**+ satisfaction with a new piece of information; **m**− frustration and shaded box; last column, final intellectual satisfaction, Likert scale from 1 (low) to 4 (high)

In contrast, on the right of the diagonal, we observe crucial questions only (Table 13.1) and no expression of agreement and much frustration (Table 13.2):

- B, text 4: It poses problem more than it explains.
- A, text 5: Err, in fact it doesn't really explain.

These findings suggest that most students needed to reach a threshold of comprehension *beyond mere logical necessity* before activating their critical potential. Once this (student-dependent) threshold was reached, agreement, moderate satisfaction and anecdotal questions disappeared, to be replaced by frustration, crucial questions, critiques (including self-critiques) and an active search for comprehension until, finally, the student was satisfied with the last explanation. We designated this intellectual dynamic as *delayed critique.*

We also observed that two students who knew the topic very well were satisfied both with their own responses and with all the presented explanations (however incomplete). We described this form of critical passivity as *expert anaesthesia.*

Hot Air Balloon

The above findings align with previous results related to the topic of hot air balloons (Mathé and Viennot 2009). One typical teaching ritual presents hot air balloons as isobaric when calculating the internal temperature needed for take-off (see, for instance, Giancoli 2005). Quasi-unanimously, teachers offer no objection to this hypothesis ($N = 129/130$, Viennot 2014b), even though their knowledge of this domain should tell them otherwise. Specifically, they should know that an isobaric situation inevitably results in a crash, as all kinds of flotation link to a gradient of pressure or that the same pressure on both sides of an envelope cannot result in a force exerted on this envelope.

For the purposes of this study, 14 prospective journalists (in their third year at university) were asked for their opinions of an article explaining how a hot air balloon works. This mentioned the inconsistent hypothesis referred to above, and while interviewees realised at various moments that the hypothesis was absurd, there was in most cases a noticeable delay in explicitly criticising the article itself. This result can now be understood as a case of delayed critique.

Survival Blanket

A subsequent investigation revealed another intellectual dynamic of interest. A small group ($N = 7$) of prospective teachers (in their fourth year at university) were interviewed about how to use a survival blanket for protection against the cold (Viennot and Décamp 2016a). Here again, a delayed critique dynamic was observed. In this case, students' judgments were strongly influenced by a preconception: that the best possible way to protect against cold with a survival blanket was to ensure the maximum reflection of "heat" towards the body. Not surprisingly, then, participants were found to have difficulty in critiquing texts that presented the same view. In the case of one participant, however, an element of information available from the start (that the gold side is more emissive than the silver side) was used to trigger an "early critique" so confirming that this dynamic was logically possible, although the other participants seemed unable to access it.

13.4 Recapitulation and Discussion

The three experiments described above supported a first recapitulation of our findings (Table 13.3).

These results—recently completed by a study on the topic of osmosis (Viennot and Décamp 2016b)—support Willingham's view that "Critical thinking is not a set of skills that can be deployed at any time, in any context" (Willingham 2008). To put it in another way, students' critical development cannot be seen as independent of their conceptual progress. While advanced students' responses to incomplete or incoherent explanations show that they are searching for coherence, most of them need to reach a threshold of comprehension *beyond logical necessity* before expressing their intellectual frustration—a symptom of delayed critique. In rare

Table 13.3 Numbers of interviewees manifesting each of the three identified intellectual dynamics in the three investigations summed up above

	Sample	Delayed critique	Early critique	Expert anaesthesia
Topic				
Authors				
Isobaric' hot air balloon (Mathé and Viennot 2009)	14 Future science journalists	12	2	0
Radiocarbon dating: Constancy of $^{14}C/^{12}C$ ratio in the atmosphere (Décamp and Viennot 2015)	10 Student teachers (physics)	8	0	2
Survival blanket: 'Put the silver side inside to protect against cold' (Viennot and Décamp 2016a)	7 Student teachers (physics)	6	1	0

The diagrams in col. 3 to 5 symbolise the respective evolution of conceptual understanding (above) and critical attitudes (below) in students during the interview

cases, this dynamic is observed after a very short delay (early critique). We have also observed that some students or teachers exhibit strong critical passivity towards certain defective explanations, even when they understand the topic very well; we call this (absent) intellectual dynamic *expert anaesthesia*. This syndrome is a good candidate to explain (at least in part) the remarkable stability of some teaching rituals (Viennot and Décamp 2016b), which can scarcely be ascribed only to habit.

It is striking that the above studies illuminate two distinct intellectual dynamics—delayed critique and expert anaesthesia—that are both marked by strong or total critical passivity. In the interpretations that (hypothetically) present themselves, conceptual development seems to play two opposing roles. In the case of delayed critique, conceptual comprehension is defective, and the respondent's energy seems mobilised by a search for previous knowledge and a desire to understand the topic before offering any critical judgment. In the second case, experts' sense of competence seems to block any concerns about the presented explanation, and it seems possible that they may unconsciously complete what is missing in the given text or drawing. Two circumstances are likely to aggravate this situation: when a common idea or echo-explanation affects experts' judgment or when a faulty explanation leads to a correct answer (see, for instance, Kahneman 2012, 52).

The identification of varying conceptual and critical intellectual dynamics may facilitate decisions about teaching goals and strategies, as well as about teacher formation. For instance, the case of a student teacher who recognises the incompleteness of an explanation but fails to draw any firm conclusion is worthy of attention. Such cases of "delayed critique" designate a target—to express one's frustration even in the absence of complete comprehension. It seems likely that intellectual dynamics such as these have important implications for how the individual appropriates teaching documents or popularised resources.

In any case, these findings serve to clarify the role of metacognitive-critical-affective factors in students' intellectual dynamics. Metacognition and psycho-affective factors have long been thought to correlate with student cognitive performance (e.g. Pintrich et al. 1993; Rhöneck et al. 1998; Launkenmann et al. 2003). Here, it is important to note the focus on intellectual processes rather than on correlations; for instance, the two opposing roles of conceptual expertise in our findings would, if further confirmed, invalidate any conclusion that relied on mere correlation with a given psycho-cognitive factor. More precisely, more attention should be paid to the possible *interconnections* between conceptual and metacognitive-critical-affective awareness. Fine-grained analyses of the kind described here seem essential in meeting the clearly urgent need to incorporate critical development in the design of teacher formation.

References

Bandura, A. (2001). Social cognitive theory: An agentic perspective. *Annual review of psychology*, 52(1), 1–26.

Décamp, N. & Viennot, L. (2015). Co-development of conceptual understanding and critical attitude: analysing texts on radiocarbon dating. *International Journal of Science Education,* 37 2038–63.

European Commission, 2015. *Science education for responsible citizenship,* Report EUR 26893 EN, Brussels

Giancoli, D.C. (2005). *Physics (6th ed): Instructor Resource Center* CD-ROM, Prentice Hall.

Jenkins, E. W. (2007). School science: A questionable construct? Journal of Curriculum Studies, 39, 265–282. doi:https://doi.org/10.1080/00220270701245295.

Jimenez-Aleixandre, M. P., & Puig, B. (2012). Argumentation, evidence evaluation and critical thinking. In B.J. Fraser, K. Tobin, & C. McRobbie (Eds.), *Second international handbook of science education* (pp. 1001–1015). Dordrecht: Springer.

Kahneman, D. (2012) *Thinking Fast and Slow*. London Penguin books.

Kattmann U. & Duit R. (1998): The model of educational reconstruction. Bringing together issues of scientific clarification and students' conceptions. In Bayrhuber B(ed): What-Why-How? *Research in Didaktik of biology*, 253–262

Komorek, M. & Duit, R. (2004). The teaching experiment as a powerful method to develop and evaluate teaching and learning sequences in the domain of non-linear systems. International Journal of Science Education, 26(5), 619–633.

Laukenmann, M., Bleicher, M., Fuller, S., Gläser-Zikuda, M., Mayring, P., & Rhöneck, C. V. (2003). An investigation of the influence of emotional factors on learning in physics instruction. *International Journal of Science Education*, 25(4), 489–507.

Mathé, S., & Viennot, L. 2009. Stressing the coherence of physics: Students journalists' and science mediators' reactions, *Problems of education in the 21st century*. 11 (11), 104–128.

Ogborn, J. (1997). Constructivist metaphors of learning science. Science & Education, 6, 121–133. doi:https://doi.org/10.1023/A:1008642412858

Pintrich, P. R., Marx, R. W., & Boyle, R. A. (1993). Beyond cold conceptual change: The role of motivational beliefs and classroom contextual factors in the process of conceptual change. *Review of Educational Research*, 63(2),167–199.

Rhöneck, C. V., Grob, K., Schnaitmann, G. W., & Völker, B. (1998). Learning in basic electricity: how do motivation, cognitive and classroom climate factors influence achievement in physics? *International Journal of Science Education*, 20(5), pp. 551-565.

Viennot, L. 2003. *Teaching physics*. Dordrecht: Kluwer Ac. Pub.

Viennot L. 2006. Teaching rituals and students' intellectual satisfaction, *Phys. Educ.* 41, 400-408.

Viennot, L. 2014a. *Codevelopment of conceptual understanding and critical attitude: an essential condition for physics learning*. Invited address. Frontiers of fundamental physics FFP14, Marseille July 2014. PoS(FFP14)011 http://webcast.in2p3.fr/videos-ffp14_coevelopment_of_conceptual_understanding_and_critical_attitudean_essential_condition_for_physics_learning_laurence_viennot

Viennot, L. 2014b. *Thinking in physics The pleasure of reasoning and understanding* Dordrecht: Springer/Grenoble Sciences.

Viennot, L. 2015a. Spotlighting a content for teaching: research examples at university level, plenary address, *L'esperienza del PLS: guardando oltre* —Roma, 11-12 maggio 2015, text on request from the author

Viennot, L. 2015b. Thinking the content for physics education research and practice, in Fazio C. & Sperandeo-Mineo R.M. (Eds.) *Teaching/Learning Physics. Integrating Research into Practice,* Proceedings of the GIREP/MPTL 2014 International Conference, Università degli Studi di Palermo, ISBN 978-88-907460-7-9, pp. 61–79 - http://www1.unipa.it/girep2014/proceedings/GIREP-MPTL%202014%20Conference%20Proceedings.pdf

Viennot, L. 2016. The persistence of Teaching Rituals, *Physics Education* 51 030104

Viennot, L. & De Hosson, C. 2015. From a Subtractive to Multiplicative Approach, A Concept-driven Interactive Pathway on the Selective Absorption of Light, *International Journal of Science Education*, 37:1, 1-30. doi: https://doi.org/10.1080/09500693.2014.950186

Viennot & Décamp, L. 2016a. Co-development of conceptual understanding and critical attitude: toward a systemic analysis of the survival blanket, *European Journal of Physics*, 37 doi:https://doi.org/10.1088/0143-0807/37/1/015702

Viennot & Décamp, L. 2016b. Conceptual and critical development in student teachers: First steps towards an integrated comprehension of osmosis, *International Journal of Science Education*, https://doi.org/10.1080/09500693.2016.1230793

Willingham, D. T., 2008 Critical thinking: why is it so hard to teach? *Arts Educ. Policy Rev.* 109, 21–32.

Part IV
Diversity and Difference in Teaching Physics

Chapter 14
Indigenous and Afro Knowledge in Science Education: Dialogues and Conflicts

Antonia Candela and Johanna Rey

Abstract This chapter provides ethnographic descriptions and analyses of interviews with indigenous and Afro-Colombian (The term refers to the descendants of Africans who survived the slave trade and to their dual affiliation: to both their black African roots and the Colombian nation. In some articles, especially those from Africa, the original African cultures are called "indigenous" (Semali and Kincheloe (Eds), What is indigenous knowledge? Voices from the academy. New York and London: Falmer Press, 1999). However, in America they are called "Afro" in order to distinguish them from the original American cultures.) teachers and of some discursive interactions with their students in primary school classrooms in underserved communities. In those contexts they mobilize their local community knowledge for science lessons. We analyzed the teachers' purpose in incorporating indigenous and Afro knowledge in teaching science and how these different knowledge systems work in the interaction. These teachers' and students' co-constructions modify and enhance the official science curriculum with forms of resistance to the scientific myth of only one universal truth about physical phenomena. This resistance is based on the strength of their collective identity constructs as well as their connection with and respect toward nature. These kinds of studies are relevant references for a culturally sensitive science curriculum development.

14.1 Introduction

Historically, Western education in neoliberal countries reproduces Eurocentric (a form of ethnocentrism) science proposals as the only legitimate and true kind of knowledge about the physical world. Neoliberal countries tend to homogenize educational policies that, aside from propping up the interests of the global market

A. Candela (✉) · J. Rey
Center of Research and Advanced Studies, Ciudad de México, CDMX, Mexico
e-mail: acandela@cinvestav.mx

M. Pietrocola (ed.), *Upgrading Physics Education to Meet the Needs of Society*,
https://doi.org/10.1007/978-3-319-96163-7_14

(Valenzuela 2003), encourage a relationship of control and exploitation of nature. However, if we understand that knowledge is situated as a product of activity in a cultural context (Lave 2011), we have to accept that different systems of knowledge are developed from the cultural diversity of the schooling participant (Colbern and Aikenhead 2003; Barnhardt and Kawagley 2005; McKinley 2007).

In the face of these trends, we are interested in studying whether teachers, especially from indigenous and Afro backgrounds, resist educational proposals that do not take into consideration the students' and communities' cultural worldviews.

In this paper we assume that any sociocultural configuration, such as school, is a product of a variety of global, national, regional, and local traditions that are interrelated in ways that on the one hand reproduce the neoliberal system but on the other construct spaces where alternative practices and knowledge are cultivated, inasmuch as these traditions have differing degrees of relative autonomy with respect to global proposals (Rockwell 2009). It must be recognized, however, that the different traditions have both similarities and differences, which can lead to internal contradictions when it comes to day-to-day teaching practice. These contradictions produce developments, changes, and transformations in this practice, a dialectical relationship of reproduction/resistance in response to the proposals of the dominant system. In multicultural countries of Latin America, such as Mexico and Colombia,[1] some of these traditions and influences are imposed by the power of the national system, but others have their roots in alternative historical contexts situated in local networks, in oral traditions, and even in pre-Hispanic heritage (Rockwell 2009), primarily in indigenous and Afro-descendent communities that preserve representative features of their ancestral culture (Bonfil 1990).

As certain ethnographic studies show, teachers and students come to the classroom with knowledge and practices from the local culture, thus constructing a diversity of everyday school cultures (Rockwell 1997). In line with these findings, Nespor (1994) contends that classrooms have permeable walls that allow for communication with other spaces and times, specifically the culture of the community where the schools are located.

This article seeks to analyze empirical data from classroom interviews and interactions in order to find out how science is taught in schools situated in indigenous contexts in Mexico and in Afro-descendent contexts in Colombia. We conducted ethnographic research to see whether different knowledge systems coexist in these educational contexts and to understand how these divergent systems are managed in everyday teaching practice. The aim is to describe the relationship between the knowledge systems rooted in Western science and promoted by national

[1]In Mexico, more than 68 ethnic groups (8% of the whole population) keep their culture alive through their language and traditions. Colombia also has a rich ethnic diversity with 94 indigenous groups maintaining 64 different languages. 3.4% of the Colombian population is indigenous and 10.6% is Afro-Colombian. These groups are marginalized and tend to live in the worst economic and social conditions within the neoliberal system of the region, with underserved schools as a consequence (Walsh 2007).

study plans and the alternative knowledge systems based on different cultural referents that teachers and students from indigenous and Afro-descendent cultures bring to classroom interaction.

The importance of an ethnographic study such as this one lies in the potential to provide information from the perspective of educational actors, in order to provide information to attend the growing demand of different ethnic groups around the world for a democratic education that takes its knowledge into account, especially as it bears on their relationship with the natural world. Most of the movements undertaken by the 50 million indigenous inhabitants of Latin America, such as the Zapatista movement, call for and try to develop an education that respects their identity-affirming conceptions and their vision of nature (López 2001). Of particular interest is determining whether participants bring to science classrooms their ancestral ethnic group's relationship with nature, which over millennia has proven its ability to maintain sustainable development that ensures the survival of both humans and other species. As Barnhardt and Kawagley (2005) argue, the ancestral knowledge of original groups can benefit all students as it upholds values, beliefs, and practices that are increasingly being recognized as legitimate and relevant for today's world.

It must also be taken into account that globalization generates a growing inequality that leads to the migration of large human groups, who have to leave their cultural context of origin. This situation contributes to a great presence of multicultural classrooms in schools of most countries all around the world, many of them in need of intercultural education (IE) in order for their students to better understand and learn. Even though a significant number of intercultural proposals have been made around the world, especially in non-developed countries, there are less studies that analyze empirical data of the procedures that teachers and students from indigenous and Afro communities use in including their local knowledge in science education. There are also few articles that study the indigenous or Afro knowledge topics that are spontaneously included in those classroom contexts and the teachers' motivations to do it.

Understanding the dialogues, conflicts, and constructions between indigenous and scientific systems of knowledge in classroom interaction (Bang and Marin 2015) and how educational participants deal with them can contribute to further intercultural proposals for science education as a task in construction (Godenzzi 1996) that can challenge the Eurocentric, positivist perspectives of science education.

In this sense, this article analyzes voices (Bakhtin 1981) from teachers and students in indigenous and Afro-Colombian communities who incorporate local interpretations of the natural environment into their classroom interaction, despite operating educationally within national Eurocentric science programs that do not consider alternative conceptions of nature.

This paper's results show teachers from indigenous and Afro-Colombian backgrounds manifesting ethical responsibilities to nurture scientific knowledge with the cultural perspective of their original communities, particularly about nature. They all challenge the irrational exploitation of nature that Western scientific anthropocentric approaches promote by mobilizing their spiritual and holistic relationship with the natural world in their classrooms. This can be seen as a form of resistance against scientific knowledge as the sole and universal truth. The empirical data of this paper bring to light the tensions between the production of scientific and indigenous knowledge in these communities, as challenges to the cultural reproduction of Western scientific perspectives. These results bring into debate what kind of voices and knowledge production these educational actors can reconcile and what others prove to be resistant to negotiation. The paper offers empirical examples of what educational practices must take into account when working with diverse cultural knowledge constructions without denying or jeopardizing any of them.

In order to develop the argument of the paper, we describe academic research that studies the problems of dialogue between scientific knowledge and other cultural knowledge constructions (Godenzzi 1996). We also present some theoretical advances of science education orientations that discuss the possible coexistence in the same person, organization, or community of divergent systems of knowledge about the physical world (Barnhardt and Kawagley 2005). We describe features of the fieldwork and methodological approaches to analyze the collected empirical data and arrive at some final thoughts.

14.2 Indigenous and Scientific Knowledge

Aikenhead and Owaga (2007) review the origin and features of scientific knowledge and of indigenous systems of knowledge in order to explain similarities and differences between those systems of knowledge that allow for building some bridges between them. These authors show that both systems are culture-dependent and that neither is superior to the other, only more pertinent depending on the context. They both exhibit rational thinking, are predictive, use empirical approaches, and are continually being revised in the light of new observations and the contributions of other conceptions. However, they also have relevant differences. Indigenous knowledge tries to be harmonious with nature, while science sets out to dominate it. Indigenous knowledge is monistic because it does not separate matter and mind and sees everything in the universe as alive: animals, plants, humans, rocks, celestial bodies, natural forces, etc. (Cajete 2000). In indigenous knowledge, inner space (the spiritual world) and outer space (the physical world) interact holistically (Ermine 1995). In Eurocentric science, on the other hand, Cartesian dualism separates mind and matter, and the connection to nature tends to be "reduced to a relationship of material production" (De Sousa 2015: 15).

However, these different knowledge systems are constantly adapting and changing in response to new conditions and in relation to their interactions (Barnhardt and Kawagley 2005). Even Western science incorporated knowledge from different cultures such as Chinese, Arabian, and Mesoamerican.

In the literature there is a debate regarding how to relate this conceptual and epistemological diversity for science education proposals. Pomeroy (1992) summarizes this debate by postulating nine research agendas for teaching science among different cultural systems; these agendas can be grouped into two different perspectives: (1) assimilationist, all those proposals that take into account indigenous themes, explanations, and/or languages in order to include them in a scientific approach by trying to explain them from a scientific perspective, emphasizing Western science's power to explain and predict, and (2) anthropological or autonomous enculturation proposals, those that compare and build bridges and analyze epistemological conceptions and explanations of the physical world contributed by Western science as well as everyday and other cultural systems of knowledge, with no attempt to subordinate or eliminate any of them.

June George (1999) provides interesting ideas for using indigenous knowledge as a component of school science curriculum, developed from his work in Trinidad and Tobago. He constructs four categories in order to build bridges between science and indigenous knowledge: Category 1. Indigenous knowledge can be explained in conventional science terms. Category 2. A conventional science explanation for indigenous knowledge seems likely but is not yet available. Category 3. A conventional science link can be established with the indigenous knowledge, but the underlying principles are different. Category 4. Indigenous knowledge cannot be explained in conventional scientific terms.

George suggests the use of indigenous knowledge from category 1 in science classes in order to highlight similarities between the two systems, generate interest among the students, and develop pride in the knowledge and wisdom of their ancestors. Knowledge from category 3 can also be a good point of departure to explore and discuss their resemblance. For George category 2 is a fertile area for scientific research; however, constructing relations between the two systems of knowledge is frequently beyond the students' capabilities. I think this kind of knowledge can develop identity values and pride for indigenous students as well as respect and some interesting ideas for other students to learn. Knowledge from category 4 represents the biggest challenge for the teachers. For this last kind of knowledge, he suggests exposing the students to both knowledge systems to illustrate that in the conduct of our lives, we sometimes draw on different knowledge systems. This last idea seems to take into account the possibility of addressing diverse knowledge systems in order to use them in different contexts.

Aikenhead and Owaga (2007) propose recovering and continuing George's initial work in order to classify more topics of indigenous knowledge and to study how they can be used in science classes. However, they take note of the epistemological differences that must not be overlooked. For example, they describe that in holistic indigenous thought (which does not separate mind and matter), a herbal cure cannot achieve the same effect without the ceremony and ritual songs, chants, prayers, and relationships they used to have in the original context (Battiste and Henderson 2000: 43), even though it can also be explained in scientific terms.

For Bang and Marin (2015), science education that takes into account indigenous systems of knowledge improves the quality of learning and opens opportunities for students from historically non-dominant communities. Building these science learning environments expands the boundaries of reality and possible futures for students.

14.3 Science Education Approaches

The proposals for science education have been changing as a result of theoretical and epistemological developments since the 1960s.

Recent advances in educational research come from connections among different fields and research traditions such as psychology, sociology, anthropology, and the research results of studies of language in sociolinguistics, linguistic anthropology, and diverse discourse analyses. An important task is to articulate related fields and confront and complement diverse disciplinary traditions and approaches in order to understand classroom dynamics where formal education takes place. In order to capture the complex processes occurring in classrooms, research must necessarily be interdisciplinary (Candela et al. 2004).

Some philosophical and sociological approaches help to redefine the epistemological status of science as a cultural construction (Elkana 1982) that cannot sustain its superiority over other cultural systems of knowledge, only its more pertinent explanation in some contexts.

One of the most influential models based on the Piagetian psychogenetic perspective was the conceptual change model (Posner et al. 1982) that develops some procedures in order to generate a cognitive conflict with "misconceptions" (ideas that differ from scientific postulates) in order to eliminate them and construct scientific concepts. This model is based on the idea that scientific concepts are incompatible with "common sense" or other cultural ideas.

Challenging the model proposed by Posner et al. (1982) of changing everyday ideas for scientific ones, two decades of research have shown that students, even from university level and after several science courses, can still have "common sense" ideas about some scientific topics. The responsibility for this problem has frequently been attributed to pedagogical deficiencies or to teachers' incompetence. Efforts to improve teaching models have not had significant results in eliminating "common sense" and other cultural ideas.

To deal with this problem, one recent and productive line of research in science education confirms that everyday ideas can coexist with scientific conceptions (Scott 1987). Relevant contributions in this sense have been made by Mortimer (1995), who questioned the ontological and epistemological backgrounds of the conceptual change model because it is based in the empiricist idea that people have a single conceptualization coming from a direct perception of the natural world. Mortimer states that everyday ideas do not disappear when a person appropriates other notions about the same topic.

Today science education perspectives recognize that several cultural, common sense, and scientific ideas can coexist, even in adults with an academic education, as they are useful in some contexts. This is what happens with religious and magical ideas that even well-known physicists maintain for everyday, psychological, or emotional needs. It is also important to take into account that these different conceptions are not necessarily coherent with each other and can be held without generating personal conflicts (Hodson 1999).

Taking into account the possible coexistence of different cultural systems of interpretation of the physical world, advanced developments of science teaching

(Mortimer 1995) propose helping students develop the ability to analyze different conceptions of the world and to make decisions on the basis of the most pertinent perspective and explanation in each concrete situation.

Another important influence of psychology on science education came from the sociocultural approaches grounded in Vygotskian ideas (1984). For Vygotsky knowledge construction was based not only on individual relationships with natural endeavors, as the Piagetian perspective states, but also on the cultural and historical conceptions interiorized from social interaction in order to interpret the natural environment. These developments contribute to understanding why students from different cultural backgrounds can have divergent interpretations of the physical world.

14.4 Fieldwork and Empirical References

This is a qualitative (ethnographic) research paper that sets out to analyze the meaning of teaching ideas for educational actors (Erickson 1986). In the analysis of the interviews and the excerpts of interaction in the classrooms, we understand that educational actors do not only interact "face to face" among themselves. We assume they interact simultaneously with people, cultural artifacts, and diverse representations that they mobilize from distant spaces and times (Nespor 1994).

The paper provides sociocultural analysis of fieldwork notes and audio recordings of semi-structured interviews with a Mexican indigenous (Tere) and an Afro-Colombian (Stella) science teacher in the context of a Western science curriculum for their underserved primary schools in indigenous and Afro-Colombian communities. Other data came from five semi-structured interviews about the experience of a physics teacher (Juan) after 2 years of teaching in an intercultural program at an indigenous high school (Candela 2013). We also analyze some extracts of 3 audio recordings of Tere (with 19 children from combined first- and second-grade classroom) and 48 audio videos of Stella's science lessons (fifth grade) that open up alternative spaces by mobilizing ancestral knowledge from their communities.

We became familiar with these contexts in relatively time-extended stays for the purpose of building specific knowledge through the documentation of participants' discourse and practice. We presented ourselves to the teachers as researchers interested in analyzing how they teach science in the context of indigenous and Afro-Colombian communities.

Tere was born in an indigenous Purépecha community of the Mexican state of Michoacan. She studied an undergraduate degree in Pedagogy and two master's degrees—one in Learning Difficulties and the other in Psychogenetics. She has 29 years of teaching experience in one-room elementary schools (teaching six grades simultaneously) in indigenous communities, as well as at graded monolingual schools, both Purépecha- and Spanish-speaking. She was working at a K-6 school in Ichupio, a community on the banks of Lake Pátzcuaro in Michoacan. Ichupio has approximately 350 inhabitants, most of them Purépecha-speaking, who earn their living by fishing and by producing and selling agricultural and handcrafted products, activities that generally bring in little income. This teacher has been working with some other indigenous teachers in the construction of culturally sensitive teaching proposals for their region. The neoliberal national educational program imposed by the Mexican government does not recognize this kind of experience, however, and they cannot control everyday practices at the schools.

Stella is a Colombian teacher who was born in Cali and identifies herself as *Afro-Colombian*. She has undergraduate degrees in Biology and Chemistry; a master's degree and specialization in Ecology, Environment, and Development; and experience in Afro ethnoeducation and the teaching of Afro-Colombian studies. She has over 30 years of teaching experience, both in public and private schools and at the kindergarten, elementary, middle, and high school levels, in both urban and rural contexts. She works at a school in southeastern Bogota, a district known for low-income levels and precarious living conditions. Many families are immigrant or displaced who have come to the city looking for job opportunities and decent living conditions; job opportunities, however, are scarce. The ethnographical data were collected in a group of 41 fifth and sixth graders during several prolonged stays over a period of 2 years and 3 months. Forty-eight classroom video recordings were made as well as three audio interviews with this teacher.

The indigenous teacher interviewed in the Tzeltal community of Chilón, in the Mexican state of Chiapas—called Juan for confidential reasons—studied physics for 4 years at a local university in Chiapas. He has 15 years of experience teaching this discipline at the high school level in Tzeltal communities. He shares the culture and language of his 20 Tzeltal students. The Tzeltals are one of the indigenous groups that still have their own language and culture in Chiapas. Chilón has 395 inhabitants, 84 of whom are monolingual in the Tzeltal language and the rest speak Tzeltal and Spanish. Almost the entire population works in the agricultural sector and has a low socioeconomic level. We have ethnographic notes from Chilón, where we stayed for 2 weeks, and five interviews with Juan at the school.

After transcribing all the interviews and classroom recordings, and reading them a number of times, we selected discursive excerpts from the three contexts in which we found relevant information about the relationship between indigenous or Afro-Colombian and scientific systems of knowledge. In the following section, we analyze the arguments the teachers provide for including local knowledge in their science lessons. Further sections present analysis of the way they deal with both systems of knowledge in classroom interaction. These excerpts were analyzed in an attempt to understand the significance of the natural world that the participants construct within their local knowledge and the scientific worldview.

14.5 Self-Recognition of Their Own Culture

In this section we analyze excerpts of interviews with the teachers from the three contexts (Tzeltal and Purépecha communities of Mexico and one Afro-Colombian community), in which they advance some arguments about why they consider local knowledge when teaching science and how they do it.

In the following discursive sequence, Juan not only confirms that he mobilizes Tzeltal cultural knowledge in the physics lessons, but he also explains to us why it is important to do so.

> **Juan:** We have to know about our history, our culture... our elders, so that we know what we have and appreciate who we are . . . it's a matter of raising awareness, of acting conscientiously, so that we recognize and don't forget that we have knowledge, not only knowledge but also values, and that these values can also be transmitted through the topics of these disciplines . . . but we don't have to accept everything from scientific culture . . . there are things that make life easier but also things that affect us....

> **Juan:** Tenemos que conocer sobre nuestra historia, nuestra cultura, de nuestros mayores, para conocer lo que tenemos, y valorar quiénes somos . . .es un asunto de concientización, de actuar con conciencia para reconocer y no olvidar que nosotros tenemos conocimiento, pero no sólo conocimiento sino valores y que esos valores también se pueden transmitir a través de los temas de estas disciplinas . . . tampoco tenemos que aceptar to::::do lo de la cultura científica . . . hay cosas que nos facilitan pero también hay cosas que nos afectan . . .

The first argument this teacher provides to justify incorporating his indigenous knowledge in science lessons is about the importance of preserving his indigenous culture as "acting conscientiously," raising awareness in the students about their culture and their responsibility and ethical position with regard to it. He mentions that they, referring to himself and the members of his ethnic group, must not "forget" and need to appreciate what they have and who they are. In his proud declaration of his Tzeltal identity, he also takes their history and culture through the elders' voices as a source of knowledge.

He mentions, as an implicit form of conflict with science, that they have not only knowledge but also values, a holistic conceptualization that we usually do not see in science. He points out that values come from their traditional culture. However, trying to avoid a confrontational position and constructing bridges with science, he mentions that these values can also be introduced into science lessons. It can be said that Juan revisits scientific knowledge from the perspective of the Tzeltal culture in order to incorporate some values into science. This is what Colbern and Aikenhead (2003) called "autonomous enculturation."

He adds that not everything proposed by science should be accepted. This implies that his adherence to the "mother" culture is apparently unconditional, while science's contributions are conditioned by their impact on daily community life (there are things that affect us). It is interesting to note that he uses social criteria to accept some scientific formulations, and not only their relation to empirical evidence. This orientation shows some of the collective criteria that dominate in indigenous communities, as opposed to the individualistic ideology of the Western world.

In what seems to be a similar commitment with her culture, Stella talks in an interview about the recognition and visibility of Afro-Colombian culture:

> **Stella:** (...) part of the problem is that I, as a human being, was taught that I am above everything else, that I take something and exploit it and overexploit it, until it's all used up. We believe we are superior, or in other words, we have an anthropocentric worldview. But I depend on nature. In other cultures, relationships are not pyramidal; they are about communion. So I tell the kids: in terms of culture, there's indigenous, there's Romani, and there's Afro; we're going to work with Afro because it's the culture with the least presence in

schools. It is also the way to think how I can build a broader identity, less burdened with the idea that is often taught to children: "Don't mess with him because he's black."

Stella: (...) parte de problema es que se nos ha enseñado que yo, ser humano, estoy por encima, que cojo esto, lo exploto y sobre-exploto y lo acabo. Creemos que somos superiores, es decir, tenemos una visión antropocéntrica del mundo. Pero yo dependo de la naturaleza. Hay otras culturas que su relación no es piramidal, que su relación es en comunión. Entonces yo le digo a los pelados *(los niños)* : está la indígena, está la romani y la afro; vamos a trabajar la afro porque esta cultura es la que menos entra a la escuela. Es también la vía de pensar cómo construyo una identidad mucho mas amplia y menos cargada de eso que suele decírsele a los niños, "no te metas con ese porque es negrito."

In this intervention, Stella also states as one of her responsibilities the mobilization of her and her students' Afro-Colombian culture in the classroom. However, she grounds her decision in the importance of questioning the anthropocentric vision of education that assumes human superiority that justifies the exploitation of nature, thus challenging the being/doing/knowing epistemological configuration as defined by the episteme of Western modernity.

She builds for herself and her students a "broader identity" that includes Afro and other indigenous and Romani cultural elements in order to construct a non-racist identity avoiding discrimination against Afro descendants themselves. ("Don't mess with him because he's black.") Stella supports her decision to teach about the Afro culture with the argument that it is the most discriminated culture at school among other minority groups such as the Romani. With her discourse she constructs the purpose of avoiding cultural discrimination as one of the educational goals of her practice. She also talks about mobilizing ancestral epistemological alternatives, for the purpose of configuring a body of knowledge grounded in communion with nature.

In an open interview with Tere, in which she is asked how she teaches science in her community, she said that she is aware of the importance of mobilizing her Purépecha culture regarding the relationship with nature, in the classroom.

Tere: We try to show children how science and technology are exhausting our natural resources, exploiting them irrationally. (. . .) we have to consider how our ancestors thought and that is our greatest responsibility. It is up to us teachers to preserve our Purépecha culture. . . we have to rescue and strengthen certain customs. . . .

Tere: Se busca mostrar a los niños cómo la ciencia y la tecnología vienen acabando con nuestros recursos naturales, explotándolos de manera irracional (...) hay también que pensar como pensaban nuestros antepasados y esa es la gran responsabilidad que se tiene. Es que en nosotros, los maestros, está la responsabilidad de que nuestra cultura, el purépecha, se siga sosteniendo y . . .hay que rescatar y fortalecer algunas costumbres. . . .

She expresses a rejection of the irrational way science and technology have exploited natural resources but adds a reference to teachers' responsibility to mobilize their ancestors' views, to rescue and strengthen certain customs in order to keep Purépecha culture alive, and to prevent students from reproducing the irrational exploitation of nature. With her discourse, Tere is constructing a teaching role that consists of rescuing customs that seem to be fading away.

These examples show that one of the most important ideas these teachers put into consideration for teaching science to an indigenous or Afro community, even within a national curriculum of Western science, is making the students aware of the respect that indigenous and Afro cultures show toward nature. They put this cultural knowledge and attitude in opposition to science and the technological depredation of natural resources. We can say that these teachers display culturally sensitive teaching in bringing the community's cultural knowledge and relationship with nature to science lessons because it tries to maintain harmony with nature, while science sets out to dominate it (Aikenhead and Owaga 2007).

Another important aspect to be noted is that, in these cases, the teachers' stance does not correspond to individual activism only but is supported by community decisions that are shared with them. For example, in Tere's case, she designs her educational projects considering the opinion of the community, which decides at an open assembly which problems need to be addressed in school. ("We start with community planning… that is jointly agreed upon at the parent-teacher meeting according to the issues faced.") Stella also has the full support of the community's adults ("Older adults became the leading actors in this community project, because they made it possible for this type of knowledge to circulate") to mobilize community knowledge and voices at school.

These examples show the teachers' commitment to maintaining their culture and the urgency they feel about passing on the values of their ancestral cultures to their students, particularly with regard to their respectful relationship with nature. These teachers also concur in terms of discussing in the classroom concerns that are not only their own but are also shared by the community as a collective commitment, an approach that is typical of communitarian cultures. This deeply held commitment would seem to be a first necessary characteristic for teachers to try to reconcile knowledge systems rooted in science with others that grow out of the communities' ancestral cultures. It could be that this commitment shared by all three teachers helps to explain the resistance that these cultures marshal in order to preserve their identity and inherited knowledge.

In the following section, we will look at how these different systems of knowledge of the natural world are presented and mobilized and how teachers and students wield them at school.

14.6 Alternative Knowledge Systems

In order to study the tensions among different knowledge systems of the natural worldview and how teachers deal with them, we analyze interview excerpts as well as classroom interactions.

After a question about how he manages to talk about local knowledge in his physics lessons, Juan answers the interviewer:

> **Juan:** Hmmm, well, I related it to the seasons, regarding the position of the moon, the full moon, the young moon, as they say here (.2) we observe that there are things to which science says no, that is not correct; but here, from the point of view of our culture, that is how it happens, such as, for example . . .science says that you can plant at any time, but our

mother culture says that if you plant when the moon is young the tree may grow tall, but it will not give fruit, or blessings as they call it, or it will give very little, and if you plant when the moon is full even from a small tree you will obtain good fruit.

Juan: humm, bien, yo lo hice a propósito de las estaciones del año, sobre la posición de la Luna, la luna llena, la luna tierna, como le dicen por aquí (.2), nosotros observamos que hay cosas a las que la ciencia dice no, eso no es lo correcto sin embargo, aquí, desde nuestra cultura si se cumple, como por ejemplo . . . la ciencia dice que tu puedes sembrar en cualquier tiempo, pero la cultura materna dice que si siembras en luna tierna el árbol puede crecer alto pero no da frutas, allá le llaman bendiciones, o da muy pocas, y si siembras en luna llena aún de un árbol pequeño, tendrás buenas frutas.

Juan talks of contradictions between science and Tzeltal culture regarding the relation between planting and the phases of the moon. It can be noted that he addresses the local knowledge as the legitimate one, at least in their context, through an emphatic empirical argument "this is how it happens." This way, he relativizes scientific knowledge with an implicit questioning of the universality of scientific conceptions, giving voice to the community's local knowledge ("but here"). He provides his empirical experience as a local test (about having "good fruit" if you plant when the moon is full). However, at the same time, he does not attribute a universal, impersonal, and objective nature to indigenous knowledge, since he categorizes it as a cultural and local knowledge point of view ("but here, from the point of view of our culture, that is how it happens").

He places local knowledge in context and situates science as something that he seems not to totally assume. It can also be noted that with the expression of "our mother culture," the teacher connects with his culture as a collective (our) and beloved possession conceptualized as where they came from. In this case, the logic of ancient knowledge conflicts with the logic of Eurocentric science. While indigenous communities refer to lunar phases to ensure their food supply and the sustainable use of land, those who incorporate Western scientific knowledge grow crops at any time of the year, harvesting to meet market needs and disregarding the environmental implications of their practices.

In this case, the indigenous knowledge can be classified in what George calls category number 4 since it cannot be explained in conventional scientific terms. Both knowledge systems are described in the teacher's discourse, but he clarifies their tensions and irreconcilable differences, at least with regard to current scientific knowledge about planting.

In one of Stella's lessons in a fifth grade of 41 students, a boy mobilizes Afro-Colombian knowledge through the voice of Don Miguel (played by the student), a knower of this culture that has been at the school, about the respect we need to give plants in order to obtain their healing action:

B1 (Don Miguel): If we pick, say, the *hoja de Cristo* to fight illness, we must first greet it.

B2: But, how do you greet plants if they don't have ears?

G: *(she gets upset and sternly tells the children)* Well, children, what Don Miguel says is very true, since plants are living things and all living things feel and hear, even if they don't talk the way we do.

B1(Don Miguel): There are things that science cannot understand. When you enjoy the benefits of a plant, you greet it as a person; if it's in the morning you say good morning, and if it's in the afternoon you say good afternoon.

B2: That's great, Don Miguel, I never learned that in the school in the capital city.

Ao1 (Don Miguel): Si cogemos, por ejemplo, la hoja de Cristo para lograr combatir las enfermedades hay que llegar y saludarla.

Ao2: Pero, ¿cómo se saludan las plantas, si ellas no tienen oídos?

Aa: (*se enoja y con voz fuerte dice a los niños*) Bueno mis niños lo que dice Don Miguel es muy cierto porque las plantas son seres vivos y todos los seres vivos sienten y escuchan, así no hablen como nosotros.

Ao1 (Don Miguel): Hay cosas que no entiende la ciencia. Cuando uno va a obtener los beneficios de la planta, la saludamos como a cualquier persona, si es por la mañana se le dice buenos días y si es por la tarde se le dice buenas tardes.

Ao2: Qué bueno Don Miguel, en la escuela de la capital no me habían enseñado eso.

This is a debate among children in Stella's classroom about what is scientifically known (plants do not hear because they do not have ears) and Afro knowledge (plants feel and listen as any living being does). It is interesting that boy1, playing Don Miguel, disqualifies science ("there are things that science cannot understand") and legitimizes Afro knowledge, which leads the other student to question the fact that he never learned that in the capital. The girl takes a strong stance supporting Don Miguel's assessment ("is very true") about plants as living things that can hear and feel. In this way, this classroom interaction among students questions the universal validity of scientific knowledge and validates the monistic perspective of indigenous knowledge (Barnhardt and Kawagley 2005). However, the children do not seem to merely repeat the information given by Don Miguel. They show their appropriation of the cultural perspective by being able to defend it, even denying the capability of science to understand "some things."

This excerpt shows the students addressing Afro-Colombian knowledge through a participant from the community. This knowledge is not recognized by the official Colombian curriculum, as boy2 points out in the final comment. It shows that boy2 accepts the Afro-Colombian version as legitimate knowledge ("That's great Don Miguel") after the argument given by boy1 about the importance of talking with plants and after some resistance from his previous scientific ideas.

In Tere's second-grade science lesson, similar conceptions about plants' sensitivity are mobilized:

Tere: But why should we take care of them (plants)? Let's assume that the plant does not cure my headache or anything like that, why should we take care of them?

G: Because when we pick them, they also feel pain.

Tere: pero ¿por qué debemos de cuidarlas (las plantas)? Vamos a suponer que la plantita no me sirve para el dolor de cabeza ni nada de eso ¿por qué debemos de cuidarlas?

Aa: porque cuando las cortamos ellas también sienten dolor.

It is interesting to note that after a question from the teacher asking why the children have to take care of the plants, a 7-year-old girl mentions that plants can feel pain when they are cut. The teacher's question is talking about why they have to take care of plants when people cannot use them for curing certain illnesses. But the girl's answer changes the orientation of the discursive interaction from people's use of the plants to plants' feeling. This very young girl is showing local cultural knowledge more likely learned in her Purépecha community context than at school; her people's way of being, thinking, feeling, and expressing; and their particular conception of the world and life. The claim that plants feel pain establishes a connection between Purépecha and Afro-Colombian cultural knowledge, as both acknowledge that plants can feel. Stella and Tere seem to share this interpretation of plants' feeling because they do not make any connection to scientific knowledge.

We include another excerpt in which Stella explains to the interviewer the difference between Afro-Colombian and scientific knowledge, as stated by an Afro leader in the classroom:

> **Stella:** The Afro woman told the whole story of plants as they relate to witchcraft and magic. The girls especially were so excited, since they assimilated it all into their affective circle, their passions, their love interests, and their boyfriends.

> **Stella:** La mujer afro contó toda la historia de las plantas en relación con la brujería y la magia. Las peladas *(las niñas)*, sobre todo, eran re encantadas, pues ellas metían todo eso dentro de su circulo afectivo, sus pasiones, sus amores, sus novios.

The Afro-Colombian woman, called by the teacher to give her worldview to the students, mobilizes in the classroom the cultural uses that Afro communities assign to plants in rites of magic and witchcraft for curing illnesses and obtaining wishes. Stella mentions that these types of cultural knowledge fascinated the girls, because they were able to connect these plant rituals with their affective circles ("their passions, love interests and boyfriends") as cultural ways to build affective and spiritual relationships with nature.

The use of these plant rituals recalls the ideas of Battiste and Henderson (2000: 43), when they said that Eurocentric researchers may well know the name of an herbal cure and understand how it is used, but without the ceremonies and rituals, they cannot achieve the same effect. They argue that the difference between these systems of knowledge is based on the contrasts between Eurocentric reductionism and Afro-Colombian holism.

This Afro and indigenous knowledge of plants seems to be mobilized through cultural practices that tie together nature, spirituality, and ancestral culture and cannot be explained in the terms of conventional science. These excerpts are representative of indigenous knowledge characterized as number 4 by George, because they show processes where participants construct irreconcilable Western and non-Western meanings for the same concept. The students know that science affirms that plants need physical factors (such as water, light, soil, air) in order to grow. However, those scientific ideas coexist with their cultural system of knowledge that plants also feel pain and need affective and spiritual relations with human beings, knowledge which has epistemological bases that differ from those of science.

14.7 Final Thoughts

In Latin America, there are processes of staking out positions and engaging in activities and discourses that constitute anti-epistemic movements in favor of preserving ancestral knowledge from our aboriginal cultures and mobilizing it in schools (López 2001; Godenzzi 1996). Science classes, particularly in the underserved Afro-Colombian and indigenous communities we study in this paper, are meeting places where Western science is enriched or confronted with different cultural knowledge systems, especially about different relationships with nature, brought in by teachers, students, and community members.

It is noteworthy that almost all the indigenous and Afro-Colombian knowledge brought to science teaching in this paper can be classified as category 4 as proposed by George (1999). This could suggest that this teachers' selection of what content to incorporate might be driven by the objective of working with certain ideas and knowledge that they consider relevant for preserving their culture, regardless of whether or not it can be reconciled with scientific knowledge. Their concern is to form their students within the culture of their community of origin rather than to assimilate their knowledge into science.

In the interviews carried out with Mexican indigenous teachers (Tere and Juan) and an Afro-Colombian teacher, we found that all openly manifest a responsibility to nurture science lessons with the cultural knowledge of the community in which they live and teach. The commitment to preserve local culture has to do with the teachers' perception that indigenous and Afro-Colombian cultures are under threat, marginalized and excluded from the official national curricula. Tere talks about customs that must be rescued and strengthened for Purépechas to survive. Stella mentions that Afro-Colombian culture is the least present in school and talks about the importance of rejecting manifestations of racism by some students toward others because of their skin color. Juan emphasizes the importance of maintaining their history and culture, but above all, the values that science cannot give them, and allows them to discern which contributions of science to accept and which to disregard, depending on the impact on their community. These teachers come up with similar forms that can be defined as resistance when confronted with the alleged homogeneity of neoliberal globalization in different contexts—Michoacan and Chiapas in Mexico and Bogota in Colombia.

All three teachers show a commitment to their students that is not only pedagogical but also ethical, as they consider it important to promote their self-recognition as indigenous or Afro-Colombian, so that they can act with awareness of what they have and who they are, of their own histories and cultures, in what could be called the construction of their cultural identity. Contributing intentionally to the self-affirmation of one's identity by expressing appreciation in school for worldviews from non-dominant communities is an educational commitment: these teachers stand up to the powerful dominant system as a resistance to be culturally ignored advocating for the preservation of their cultures, knowledge, practices, and values. The cultural knowledge systems teachers bring to the science lessons are shared with the students, as shown in their interventions.

One of the main interests the teachers openly express is their opposition to the anthropocentric notion of science when it comes to the natural world (Stella) and the irrational exploitation of nature endorsed by the scientific and technological perspectives, which differs from the sustainability promoted by the ancestral knowledge of the indigenous and Afro cultures. In Stella and Tere's classrooms, the voices of members of their communities (Afro woman and Don Miguel) are mobilized in school in order to develop a spiritual and holistic relationship with nature. It is noteworthy that the girls are fascinated by the rituals and magical conceptions of plants, and tie this knowledge to their passions and interests, relating inner space (the spiritual world) with outer space (the physical world), as a contribution to the construction of holistic and harmonious systems between human beings and nature. These examples recall the monistic spiritual thinking behind Afro and indigenous knowledge as a process of *desettling* expectations in science education as described by Bang and Marin (2015), opening possibilities of learning for these students as shown in their increasing participation in classroom interaction. In the literature it has been noted that indigenous students feel a sense of foreignness toward science (Brandt 2007). It seems that the mobilization of their cultural knowledge could be a way to counteract indigenous students' sense of alienation.

Teachers' participation enriches science education through autonomous enculturation when they construct forms of coexistence between science and indigenous and Afro systems of knowledge without avoiding or subordinating either of them. It is perhaps an initial form of relationship among different systems of knowledge, with teachers and students interested in legitimating in the formal space of school different cultural ideas that their community has about the natural world. They seem to be asserting the legitimacy of a certain kind of ancestral cultural knowledge in school, particularly their relationship of respect for, and communion with, nature.

It is important to point out, as Aikenhead and Owaga (2007) suggest, that the teachers mobilize and establish an indigenous and Afro-Colombian knowledge that can make predictions and acknowledges an experientially grounded worldview (differences in the growth of crops planted under the full moon as opposed to the new moon), just as scientific knowledge is legitimized. In these examples, community knowledge is mobilized in schools, transforming them from being institutions that only represent the national state perspectives on knowledge to institutions that also represent the proud knowledge of the local community when the teachers convoke its cultural voices.

Following Bakhtin's work (1981), we can say that teachers' discourse creates an implicit dialogue among multiple voices: the official voices represented in national curricula, the ancestral voices of their cultures, the concrete voices of the community where they work, children's voices, and our voices as interviewers. However, when it comes to nature, much scientific knowledge is presented as irreconcilable with the knowledge of these communities, because of their very different epistemological foundations. Such is the case of the mobilization of knowledge regarding the relationship between planting cops and the phases of the moon, the sensitivity of plants, and people's spiritual and emotional relationship with them and in harmony with nature overall. They do not separate matter from mind, and thus they believe that plants have sensitivity (they hear and feel pain) (Cajete 2000).

By legitimating nonscientific perspectives on the natural world, the teachers are implicitly questioning Western scientific knowledge as the only true and universal kind. The teachers do not always propose denying scientific knowledge—at several times they insist on the importance of teaching it—but they affirm the relevance of indigenous and Afro knowledge, especially at the local level ("here, from the point of view of our culture"). By appreciating different kinds of knowledge about certain topics, they make at least a partial break with positivist science and an initial intercultural science education perspective. In this way, the analyses we present here can be seen as examples of dialogic and critical teaching of science in public schools in indigenous contexts. Each of these teachers opens opportunities to construct harmonious, holistic, relational, and complementary ancestral approaches to nature and ways of relating to it that go beyond Eurocentric scientific knowledge.

If it is accepted that schooling is a continual production and reformulation of cultural practices depending on the cultural endeavor of the institution as well as the cultural origin of teachers and students (Candela et al. 2004), then we can say that these teachers challenge and transform the epistemological perspective of the official science curricula in Mexico and Colombia when they contextualize and dispel the myth of the universality of scientific knowledge. They resist the colonial imposition of only one version of the natural world by claiming as legitimate the indigenous and Afro-Colombian knowledge that tries to be harmonious with nature (Aikenhead and Michell 2011).

The data also show that the knowledge and purposes manifested by the teachers seem to be quite similar in the two indigenous communities in Mexico and in the Afro-Colombian community, suggesting the possible existence of cultural roots shared by these ethnical communities.

However, it must be noted that all the teachers studied in this paper share a cultural background that is not limited to strict scientific formation. They share the cultural perspective of the community where they teach. This can be an explanation of their active participation in the preservation of their cultural references. Whether a culturally sensitive perspective for science education can be extended to teachers that do not have cultural backgrounds other than scientific formation is a pending research question.

Acknowledgments We thank Carol Brandt and Jrène Rahm for their comments to previous version of the manuscript.

References

Aikenhead, G. & Owaga, M. (2007). Indigenous knowledge and science revisited. Cultural Studies of Science Education (2) 539–591.

Aikenhead, G.S. & Michell, H. (2011). Bridging cultures: Indigenous and scientific ways of knowing nature. Pearson Educational Canada.

Bakhtin, M. M. (1981). The dialogic imagination: Four Essays by M.M. Bakhtin. University of Texas Press.

Bang, M & Marin. A. (2015). Nature-Culture constructs in science learning: Human/Non-human agency and intentionality. Journal of Research in Science Teaching 52(4): 530–544.
Barnhardt, R. & Kawagley, A.O. (2005). Indigenous knowledge systems and Alaska Native ways of knowing. Anthropology & Education Quarterly, 36(1), 8–23.
Battiste, M. & Henderson, J.Y. (2000). Protecting indigenous knowledge and heritage. Saskatoon, Saskatchewan: Purich Publishing.
Bonfil, G. 1990. México Profundo. Una Civilización Negada [Deep Mexico: A forbidden civilization]. México: CONACULTA/Grijalbo.
Brandt, C.B. (2007) Epistemology and temporal/spatial orders in science education: A response to Aikenhead & Owaga's: Indigenous knowledge and science revisited. Cultural Studies of Science Education (2) 599–605.
Caicedo, J. y Castillo, E. (2012). Infancias afrodescendientes: Una Mirada pedagógica y cultural. Curso para agentes educativos de educación inicial. Modalidad semipresencial. Módulo 8. Editorial Kimpres Ltda. Bogotá- Colombia.
Cajete, G. (2000). Indigenous knowledge: The Pueblo metaphor of Indigenous education. In: M. Battiste (Ed.) Reclaiming Indigenous voice and vision 181–191. Vancouver, BC: British Columbia Press.
Candela, A. (2013). Dialogue between cultures in Tzeltal teachers' cultural discourse: Co-construction of an intercultural proposal for science education. Journal of Multicultural Discourses 692–713 https://doi.org/10.1080/17447143.2012.756492
Candela, A.; Rockwell, E. & Coll, C. (2004) What in the world happens in classrooms? Qualitative classroom research. European Educational Research Journal. Vol 3 (3). http://www.uv.edu.mx/cpue/num8/contenido.html
Colbern, W., and G. Aikenhead. (2003). Cultural aspects of learning science. In: B. Fraser and K. Tobin (Eds.) International Handbook of Science Education, 39–52. Boston, MA: Kluwer Academic Publishers.
De Sousa, B. (2015). Prólogo. Boventura de Sousa Santos. En: Prácticas Otras de Conocimiento (s). Entre crisis, entre guerras. Tomo I. Cooperativa editorial Retos. San Cristóbal de las Casas, Chiapas, México.
Elkana, Y. (1982). La ciencia como sistema cultural: Una aproximación antropológica. Boletín de la Sociedad Colombiana de Epistemología No. III
Erickson, F. (1986). Qualitative Methods in research on teaching. En M. Wittrock, Handbook of research on teaching. New York: Mc Millan, p. 119–161.
Ermine, W.J. (1995). Aboriginal epistemology. In: M. Battiste & J. Barman (Eds.) First Nations education in Canada: The circle unfolds. 101–112. Vancouver Canada: University of British Columbia Press.
George, J. (1999). Indigenous knowledge as a component of the school curriculum. In What is indigenous knowledge? Voices from the academy, ed. L. Semali and J. Kincheloe, 79–94. New York and London: Falmer Press and Taylor & Francis Group.
Godenzzi, J.C. (Comp.). (1996). Educación e Interculturalidad de los Andes y la Amazonia. Perú: Estudios y Debates Regionales Andinos.
Hodson, D. (1999). Critical multiculturalism in science and technology education. In Critical multiculturalism: Rethinking multiculturalism and antiracist education. ed. S. May. 216–44. London: Falmer Press.
Lave, J. (2011). Apprenticeship in critical ethnographic practice. Chicago: University of Chicago Press.
López, L.E. (2001). La cuestión de la interculturalidad y la educación latinoamericana. VII Reunión del Comité Regional Intergubernamental del Proyecto Principal de Educación en América Latina y el Caribe ED-01/ PROMEDLAC VII
McKinley, E. (2007). Postcolonialism, indigenous students and science education. In: Handbook of research in science education, ed. S. Abell and N. Lederman, 199–226. New Jersey: Lawrence Erlbaum Associates Publishers.

Mortimer, E. (1995) Conceptual change or conceptual profile change? Science & Education 4, 267–285.
Nespor, J. (1994). Knowledge in motion: Space, time and curriculum in undergraduate physics and management. London, New York: Routledge Farmer
Pomeroy, D. 1992. Science across cultures: Building bridges between traditional western and Alaskan native sciences. In The history and philosophy of science in science education. Vol 2 Proceedings of the second international conference on the history and philosophy of science and science teaching, ed. S. Hills, 257–67. Kingston, ON: The Mathematics, Science, Technology and Teacher Education Group, Faculty of Education, Queen's Univ.
Posner, G.J., Strike, K.A., Hewson, P.W. & Gertzog, W.A. (1982). Accommodation of a scientific conception: Towards a theory of conceptual change. Science Education, 66(2), 211–227.
Rockwell, E. (1997). La dinámica cultural en la escuela. En: Hacia un curriculum cultural: La vigencia de Vygotski en la educación. Amelia Alvarez (coord.). Madrid: Infancia y Aprendizaje. Pp. 21 a 38.
Rockwell, E. (2009). La Experiencia Etnográfica: Historia y cultura en los procesos educativos. Buenos Aires: Paidós
Scott, P. (1987). The process of conceptual change in Science: A case study of the development of secondary pupil's ideas relating to matter. In: Novak, J. D. (ed) The proceedings of The Second International Seminar: Misconceptions and Educational Strategies in Science and Mathematics, Cornell University, Ithaca, Vol II, 404–419.
Semali, L. & Kincheloe, J. (Eds) (1999). What is indigenous knowledge? Voices from the academy. New York and London: Falmer Press.
Valenzuela, A., (2003) Accountability and the Privatization Agenda. In: Valenzuela, A. Leaving Children Behind: Why Texas-Style Accountability Fails Latino Youth, Albany: N.Y.: State University of New York Press.
Vygotsky, L.S. 1984. Aprendizaje y desarrollo intelectual en la edad escolar. In Infancia y Aprendizaje 27/28: 105–16.
Walsh, C. (2007). Interculturalidad y colonialidad del poder: Un pensamiento y posicionamiento otro desde la diferencia colonial", en El giro decolonial. Reflexiones para una diversidad epistémica en el capitalismo global, S. Castro-Gómez y R. Grosfoguel (eds.). Bogotá: Editorial Siglo del Hombre, p. 47,62.

Antonia Candela is a professor of the Center of Research and Advanced Studies at Mexico. Dr. Candela has a degree in Physics and a PhD in Educational Research with specialties in discourse analysis and ethnographic studies of science education. She has worked in argumentation, consensus construction, peer learning, factual construction, power in classroom interaction with focus in students' participation, and intercultural science education. She has published more than 35 articles and 30 chapters and participated in 5 national reforms as author of compulsory science programs and textbooks for primary education at Mexico. She has published three books on her research work: "*La necesidad de entender, explicar y argumentar: Los alumnos de primaria en la actividad experimental*" ; "*Ciencia en el aula: Los alumnos entre la argumentación y el consenso*" México: Paidós 1999; and, co-authoring with Gabriela Naranjo and María de la Riva, "*¿Qué crees que va a pasar? Las actividades experimentales en clases de ciencias*" México: SM/Cinvestav 2014.

Johanna Rey is a teacher of a public primary school at Bogotá, Colombia. Dr. Rey has a degree in Child Education, MSc in Education with specialty in teaching of language, PhD in Educational Research with specialties in discourse analysis and ethnographic studies of science education, and postdoctoral working on intercultural science education. She has research experience in new literacies with elementary school students and adult population, in social construction of scientific knowledge in the classroom, and in teaching of indigenous and Afro-Colombian knowledge systems in science lessons.

Chapter 15
Race, Gender, and Sexual Minorities in Physics: Hashtag Activism in Brazil

Katemari Rosa

Abstract The goal of this paper is to discuss academic climate for underrepresented groups in Brazilian physics departments. The conversation stems from looking at hate crimes happening worldwide and asking whether this hateful environment of society at large affect academic institutions. Would sexism, LGBTphobia, and racism be present in physics classrooms? Could hate speech or behavior, somehow, affect physics teaching and learning? Grounded on feminist perspectives, theories of identity, and critical race theory, the paper looks into diversity and physics education by examining the situation of race, gender, and sexual minorities in physics. Specifically, it takes on hashtag activism to analyze the experiences of students from underrepresented groups in science. The site of research is social media and the narratives produced by #MyTeacherSaid in Brazil, which was a hashtag used to reveal aggressive comments professors make to students. Results show that analyzing activism through social media can be helpful for unveiling oppressive environments in academia. Specifically, this study shows there is an oppressive climate for gender, racial, and sexual minorities Brazilian students in STEM. The comments range from subtle but harmful comments loaded with gender and race stereotypes, to open threats to students. Finally, the paper urge for a change within physics education research community to include intersectional approaches that take into account race, gender, and sexuality so that we can better understand the teaching and learning of physics, in addition to provide resources to help making more inclusive STEM environments.

15.1 Introduction

This paper is born out of a provocation made by the 2nd World Conference on Physics Education's theme, namely, "Contemporary science education and challenges in the present society: Perspectives in physics teaching and learning." The conference was held in July 2016, in São Paulo, Brazil. Initially, I was planning to

K. Rosa (✉)
Federal University of Bahia, Bahia, Brazil

M. Pietrocola (ed.), *Upgrading Physics Education to Meet the Needs of Society*,
https://doi.org/10.1007/978-3-319-96163-7_15

center my talk on issues around scientific identity development, specifically, discussing about my work on Black women physicists (Rosa and Mensah 2016). However, a month before the conference, a terrible hate crime happened, the Orlando shooting (Hunt 2016). In that occasion, a man opened fire against partygoers at a club in Orlando, Florida; the place was frequented by local gay community and friends. That same weekend, two homosexual men were found dead and carbonized inside a vehicle in a small town in Bahia, Brazil (Gauthier 2016); they were a chemistry and a physics teacher. Two weeks later, in July 2, a Black homosexual student was found dead with signs of beating and half naked at the Federal University of Rio de Janeiro campus (Martin 2016). Those events were considered hate crimes and, unfortunately, hate crimes seem to be spread all over the globe. In my perspective, these events constitute challenges our society needs to address. In this direction, this paper looks into the connections between science education and hateful academic environment, behavior, and speech.

Grounded on feminist perspectives, theories of identity, and critical race theory, this paper discusses diversity and physics education by examining the situation of race, gender, and sexual minorities in physics. Specifically, it takes on hashtag activism (Gerbaudo 2012) to analyze Brazilian educational institutions climate for underrepresented groups in science. Does this hateful environment of society at large affect academic institutions? Is sexism, LGBTphobia, and racism present in physics classrooms? Could hate speech or behavior, somehow, affect physics teaching and learning? I will use LGBTphobia to express discrimination against lesbians, gays, bisexuals, trans, queers, and other people who identify with nonheterosexual nor binary sexualities (gender nonconforming).

To address these questions, I bring elements of theoretical perspectives that help me making sense of the data analyzed in the paper, focusing on key concepts, such as intersectionality, microaggressions, and identity so that we have a common grammar for this debate. Following, there is a discussion on the role of social media for activism in the present and the presentation of data collected through #MyTeacherSaid movement in Brazil. Then, the paper moves to analyze how these hashtags portray hate behavior and speech of physics teachers and faculty toward race, gender, and sexual minorities. Finally, I suggest resources that can be used to support physics students and faculty creating a healthier climate for underrepresented groups in science.

15.2 Theoretical Perspectives

Black Feminism

The lens through which I look at race, gender, and sexual minorities in physics stems from feminist perspectives, particularly Black feminism (Crenshaw 1989). It is often said feminist movement started with suffragists, when women gathered to fight for the right to vote. It is also common to hear "women" wanted to gain public places, to work outside their houses, and to hold job positions that were just allowed to men.

However, when we look at history, we can ask ourselves who were those "women"? Black women, enslaved or not, had been working outside their homes and in heavy-duty activities for a long time; feminism, or mainstream feminism, is not talking about women of color. Sojourner Truth's speech "Ain't I a woman" from 1851 (Stanton et al. 1889) addresses this issue perfectly when she says "That man over there say that women need to be helped into carriages, and lifted over ditches, and to have the best place everywhere. Nobody ever helps me into carriages, or over mud-puddles, or gives me any best place. Ain't I a woman?". Historically, mainstream feminism has not been addressing the particularities from women of color, making it necessary for "other women" to think and fight for their rights.

Patricia Collins (2000) argues there are three dimensions of oppression that Black women have been enduring: labor exploitation, public sphere denial, and negative stereotypes. The author shows how Black women's work have been "ghettoized" to occupations related to "iron, pots, and kettles," serving to US economy and keeping those women in impoverished conditions. She says, "survival for most African-American women has been such an all-consuming activity that most have had few opportunities to do intellectual work as it has been traditionally defined" (p. 6), contributing to the stratification and segregation.

The second dimension of oppression proposed refers to forbidding Black women to vote, to have equal educational opportunities, and to equitable treatment in the criminal justice system (p. 7). The policy for keeping Black women from literacy during slavery and then later providing underfunded segregated schools and currently not providing quality education for inner city and rural people has been fundamental to keep Black women outside the political public sphere.

Finally, Collins' third dimension of oppression talks about how US society creates, disseminates, and perpetuates an image of Black women as hypersexualized, servers, welfare mothers, kitchen helpers, angry women, and other negative stereotypes. A culture that holds these images tends to dehumanize Black women, denying us a full participation in the society. Because racism and sexism became normalized within US culture, these intertwined proposed dimensions work effectively to maintain a system of oppression faced by Black women.

A departure from mainstream feminism to Black feminism places the centrality of knowledge on the lived experiences of Black women, making it a source for understanding systems of oppression that combine race, gender, and socioeconomic status. This notion of analyzing combined oppressions is known as intersectionality (Crenshaw 1993). Therefore, this framework can be helpful to look at the intersections of race, gender, and sexuality and how they can constitute a system of oppression within physics education.

Identity

The discussion through the concept of identity, in this work, is framed by ideas of practice, performance, and recognition. Identity here is understood as fluid, dependent on contexts and on how one behaves within a context (Lave and Wenger 1998).

This notion is coherent with Judith Butler's work on gender identity, in which the author argues gender is a performativity act, that is, a set of behaviors, discourse, body movements, outfits, etc., chosen to perform a desired role, consciously or not (1990).

Specifically, for the case of science, Heidi Carlone and Angela Johnson offer a framework that analyzes scientific identity formation (2007). Grounded on the experience of underrepresented minorities (URM) in science, the authors argue scientific identity is the interplay of other identities plus performance, competence, and recognition as a scientist. Therefore, to be a scientist, one has to enact social performances relevant to scientific practices, such as using tools, has to have knowledge and understanding of science content, and needs to be recognized by others as a "science person."

Through social interactions, we learn how women are supposed to behave, how we have to move, how to sit, or what we can speak. Black people learn "what type" of people we are, how to dress, and what tone to use; we also learn how a scientist is supposed to behave or look like. Therefore, for URM to be scientists, they need to engage in scientific practices and activities, perform like scientists (e.g., speaking, dressing, behaving), and be recognized as scientists – and all of those need to conciliate with all their other identities.

Critical Race Theory

The other framework used throughout this paper is critical race theory (CRT), which is a movement born out of the legal scholarship in the 1970s as a racially focused critique from Black scholars within the critical movement (Rosa and Mensah 2016). For the purpose of this paper, I will focus on two concepts present in CRT: counterstories and microaggressions.

The legal scholarship has a tradition of using storytelling to present cases; it is the fundamental form of communication in that field. The legal system tends to tell stories of (and for) dominant people. However, there are other stories to be told, the stories of silenced, marginalized, and oppressed groups. Richard Delgado calls these non-dominant narratives counterstories (1989). Counterstories do more than merely providing another perspective; they confront dominant narratives.

Chester Pierce, Jean Carew, Diane Pierce-Gonzalez, and Deborah Wills discuss the psychological effects of television on people of color and define microaggressions as the "subtle, stunning, often automatic, and non-verbal exchanges which are 'put downs' of blacks by offenders" (1977, p. 65). In his studies, Chester Pierce concluded that television would send subtle and harmful racist messages on a daily basis, constantly attacking people of color (1978). These insidious racist attacks serve to perpetuate negative stereotypes and to make people of color to learn "their place."

Although CRT focuses on race when discussing microaggressions, this concept can be extended to other forms of oppression such as class, gender, sexuality, nationality, religion, and body ability. For example, the "simple" fact of not having other female students in a physics department sends to a woman the message that the department is not her place to be; it is a male space. Similarly, at an academic environment, jokes or conversations focused on heteronormative practices or able-bodied people can constitute a microaggression for sexual minorities or people with disabilities, respectively.

Hashtag Activism

In contemporary society, the Internet has been playing an important role in our lives; online social networks have revolutionized the way we interact with one another; it has even changed the way activists have been organizing themselves. An online social platform that has been largely used by activists is Twitter. Twitter is an online social network where people can broadcast messages of 140 characters or less to a small group of friends, to a person's followers, or to the entire world. When using Twitter, and other platforms, a person can add a hashtag symbol (#) to a word, phrase, or sentence and make that message searchable by the said word, phrase, or sentence. This creates a database of all messages in the world tagged by that specific hashtag. Hashtags work like call numbers in a library system, making it easy to retrieve information if you use the right tag.

The use of hashtags in social media can produce a conversation that transcends geographical boundaries and strengths the power of broadcasted messages. When used as a form of social critique, the conversations through hashtags can create a narrative that tells the story of a specific episode or movement. As Guobin Yang exemplifies, we have already several cases of hashtag activism, like #BlackLivesMatter, #JeNeSuisPasCharlie, #IcantBreathe, #BringBackOurGirls, #StopGamerGate, #WhyIStayed, #OccupyEverywhere, #ThisIsACoup, and #MuslimsAreNotTerrorist (2016, p. 14).

Women of color are also gaining narrative agency in hashtag activism. Yang considers "narrative agency in hashtag activism as the capacity to create stories on social media by using hashtags in a way that is collective and recognized by the public" (p. 14). One example of feminism hashtag is seen in Susan Berridge and Laura Portwood-Stacer's work, when they discuss the support of women activists to violence against women in Delhi, India (2015). They show how feminists use hashtags #BoardtheBus, #StopStreetHarassment, and the #EverydaySexism project to show their support. The authors stress the potential of feminist hashtags to expose the transnational pervasiveness of gendered violence. In 2015, the journal *Feminist Media Studies* issued a special divided in three editions on feminist hashtags. Although hashtag activism keeps growing strong, it seems science educators have not yet grasped the importance of this phenomenon.

An important feature of a platform such as Twitter, especially for marginalized groups, is that it provides a space for building and sharing counternarratives and "reimagining group identities" (Bonilla and Rosa 2015). Yarimar Bonilla and Jonathan Rosa argue a hashtag can become a field site. In this paper, we use the counterstories provided by #MyTeacherSaid as a site for analyzing Brazilian academic environment for STEM students, particularly in physics.

15.3 Data Collection

The *locus* of the study is the Internet and two of its social networks, Twitter and Facebook. The object of my analysis is the narrative created by #MyProfessorSaid. I am calling #MyProfessorSaid as a collective of hashtags related to comments made by teachers and faculty and revealed online by students in Brazil. In order to capture the conversation created by #MyProfessorSaid, I have searched Twitter, Facebook, and Google for the following hashtags: #meuprofessordisse, #minhaprofesoradisse, #meuprofessorsecreto, #minhaprofessorasecreta, #esseehmeuprofessor, #esseémeuprofessor, #essaehminhaprofessora, and #essaéminhaprofessora. Currently, Google's search algorithm can retrieve hashtags from various platforms, which makes my searching approach redundant, but I wanted to be sure to search directly at Twitter and Facebook.

These hashtags translate from Portuguese as "my professor said" (#meuprofessordisse, #minhaprofesoradisse), "my secret professor" (#meuprofessorsecreto, #minhaprofessorasecreta), and "this is my professor" (#esseehmeuprofessor, #esseémeuprofessor, #essaehminhaprofessora, #essaéminhaprofessora). Because Portuguese has a gendered grammar, there is a male and a female version for each hashtag. In addition, the verb "is" can be written with accent (é) or without accent (eh), which explains four hashtags just to express "this is my professor." In Portuguese, the word "professor" can be the same for teacher, faculty, and professor. Considering the analyzed hashtags which make reference to university faculty, and that the vast majority of faculty in physics departments in Brazil is tenured, I use "professor" – both words are spelt the same in Portuguese and in English.

In addition to comments on social media, I have collected images online representing an unfolding of the same conversation #MyProfessorSaid. I have also taken pictures of posters on the wall of university campuses when I found a continuation of this conversation. A snowball approach was used since, after a while, several comments would bring the same messages I had already collected.

Race, Gender, and Sexuality: Climate in Brazilian Universities

In an article by Andrew Jacobs on July 5, 2016, *The New York Times* says "Brazil is confronting an epidemic of anti-gay violence: Despite a storied image as a tolerant, open society, Brazil is, by some counts, the world's deadliest place for sexual minorities."

Looking at Brazilian news outlets, it was not hard to quickly find these headlines:

- "After racist and homophobic protest, University of Brasília condemns offensives and will investigate the case," publishes Yara Aquino, from Agência Brasil, on June 20, 2016.
- "Assassination of Black and gay student in Rio shows intolerance at the university—Diego Vieira was found dead with signs of beaten and half naked on UFRJ campus," brings El País Brasil, on a piece by Maria Martin, published on July 7, 2016.
- "Homosexual teachers are murdered and their bodies incinerated in Bahia," writes Jorge Gauthier for Correio 24 horas, on June 14, 2016.

These headlines can give a sense of current climate for racial, gender, and sexual minorities at Brazilian universities.

Hashtag Activism in Brazilian STEM Courses

#meuprofessordisse at USP Universidade de São Paulo (USP) is the largest university in Brazil, located in the country's most populated state; it is home of our oldest physics research and teaching institution. In 2016, the Institute of Physics at USP (IFUSP) entered the world of a trending topic by hashtag feminism, #meuprofessordisse (#myprofessorsaid). This hashtag made public things said by teachers, faculty, and professors in various schools and academic settings. When IFUSP students entered the conversation, they revealed an environment of sexism, elitism, and homophobia in the country's largest physics department.

The online protest gained IFUSP physical walls, and posters were printed displaying the hashtag contents. Below are pictures of some of the messages posted on the walls.

In Fig. 15.1, it is possible to see some comments were hashtagged to indicate whether the comment was made by a male professor ("professor") or a female professor ("professora"). In this sample, comments made by female faculty expressed, specifically, forms of oppression based on race, "Do you want Black colleagues that know nothing," and class, "I am not going to change my classes to poor students who live in the slums."

There were variations in the hashtags, at USP and elsewhere, including #meuprofessordisse (my professor said), #esseemeuprofessor or #esseémeuprofessor (this is my professor), #meuprofessorsecreto (my secret professor), #minhaprofessoradisse (my female professor said), and #umdiretordisse (a dean said). Although hashtags varied, the conversation was the same, microaggressions toward women, people of color, sexual minorities, and low-income students.

#meuprofessorsecreto at UFES One feature of hashtag activism is its ability to cross geographical boundaries. At the Federal University of Espírito Santo (UFES), students also brought a narrative exposing an aggressive academic environment. Figure 15.2 shows those messages on a board; a large poster says "professor, harassment is a crime."

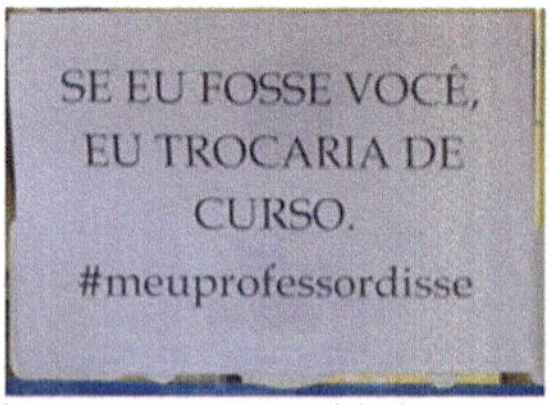

If I were you, I would change major.

Physics is for men.

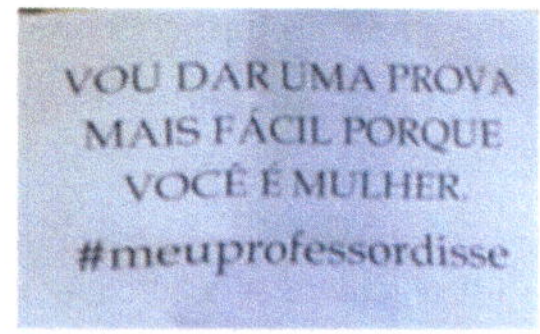

I will give you an easier exam because you are a woman.

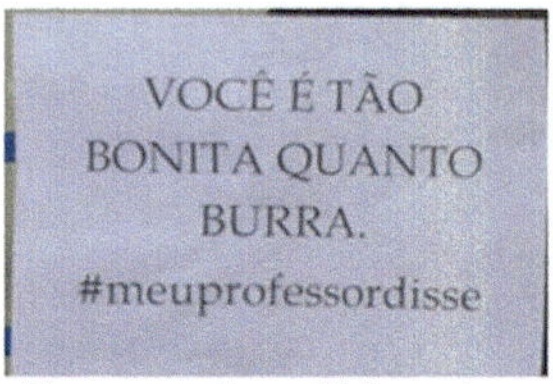

You are as beautiful as stupid.

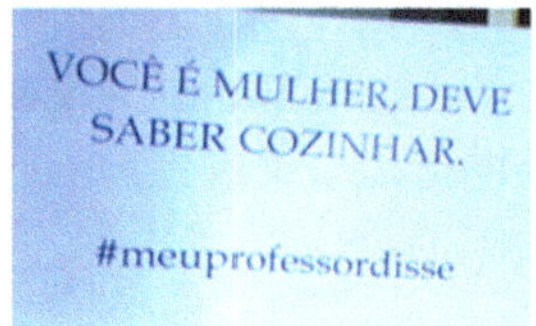

You are a woman; you should know how to cook.

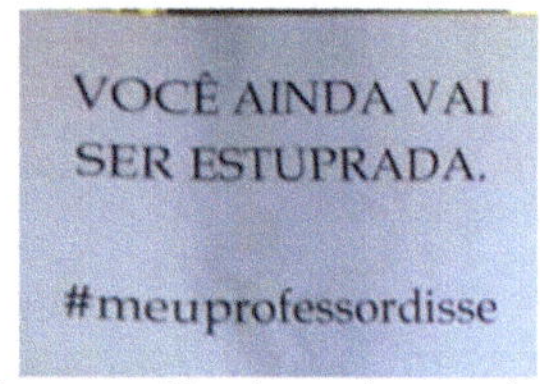

You are still going to be raped.

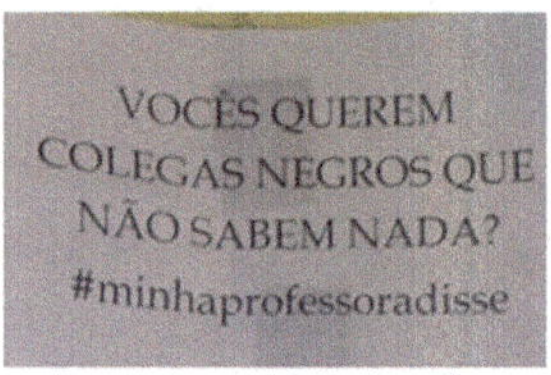

Do you want Black colleagues that know nothing?

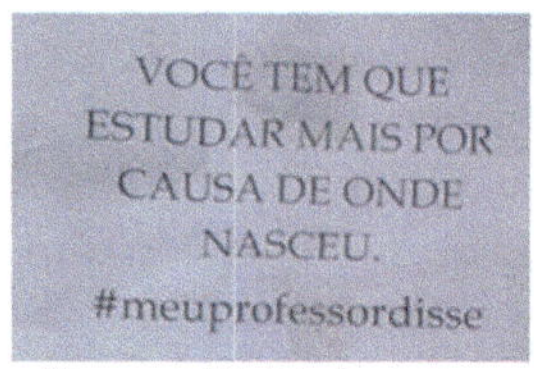

You need to study more because of where you were born.

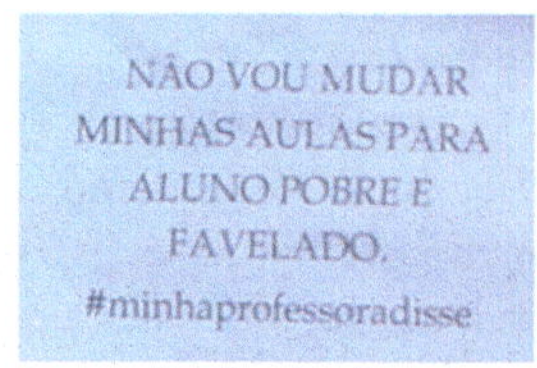

I am not going to change my classes to poor students who live in the slums.

Fig. 15.1 A sample of wall posters at University of São Paulo displaying comments made by Institute of Physics' professors to their students—#meuprofessorsecreto

Similar to USP, online activism gained physical walls at UFES. In another hand, unlike USP, where #meuprofessordisse was specifically from Institute of Physics, messages at UFES' board show the broader environment of that institution. Students' narratives through hashtags reveal harassment, objectification of women, and overall sexism (Fig. 15.3).

#meuprofessorsecreto at IFG Online activism, as wall posts show, unfolds offline. Stemming from #meuprofessorsecreto, students from Instituto Federal de Educação Ciência e Tecnologia de Goiás (IFG), an institution targeted to technology careers, wrote an open letter (Open letter 2016) asking for support regarding the abuse they have suffered at school. On the letter second paragraph, students expose sexual harassment perpetrated by IFG's faculty:

> For over 2 years we have been suffering from various abuses, there is a faculty we cannot name who is using his power to harass female students. He has even said: "If any female student would go out with him, he would pass all other students." (Open letter 2016)

Fig. 15.2 A board at the Federal University of Espírito Santo displays several sentences said by institution's faculty and denounced by students through #meuprofessorsecreto

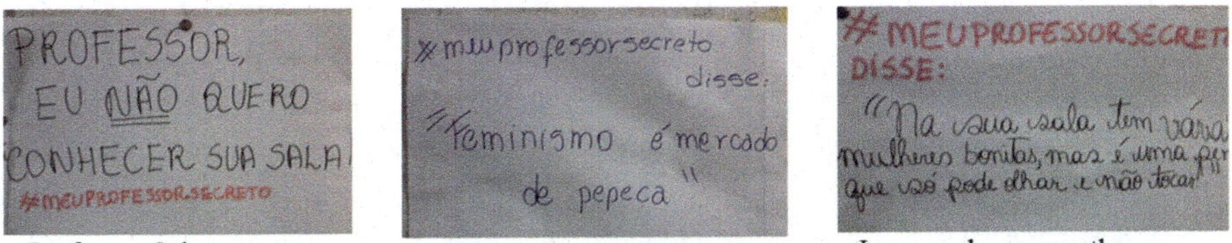

Professor, I <u>do not</u> want to visit your office.

Feminism is a [female genitalia slang] market

In your classroom, there are many pretty women, but it is unfortunate one can one look at and not touch them.

Fig. 15.3 Comments made by faculty at UFES, #meuprofessorsecreto

On the fourth paragraph, students add LGBTphobia, racism, and religion persecution as part of the academic climate at IFG:

> Several students were victims of racism, LGBTphobia, prejudice, sexual harassment, and religion intolerance. It got to the point the faculty asked: "If the student got late for being at senzala." [. . .] over 30 complaints were made against this faculty. (Open letter 2016)

Senzala is the place where enslaved people were kept captive during slavery times in Brazil.

IFG students, like students from other institutions, use hashtags in various platforms to reveal microaggressions, to reach out to other students and society at large. Figure 15.4 shows a public Facebook post using #meuprofessorsecreto; in that

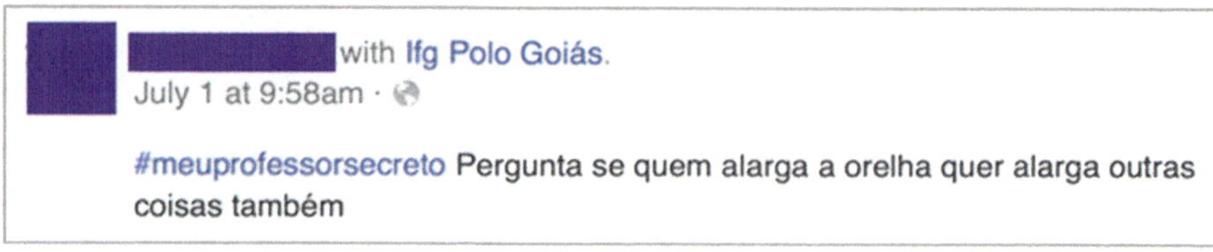

Fig. 15.4 Facebook post using #meuprofesorsecreto reveals faculty's aggressive and sexually charged comment. Translation: "mysecretprofessor Asks if those who stretch their ear are also willing to stretch other things

post, a student shares a sexually charged comment made by a faculty suggesting the use of ear stretchers, a type of body piercing jewelry, may be a sign of willingness to "stretch" other parts of the body. It is important to remember those comments are made by faculty to students at academic environments.

#esseémeuprofessor at UFRGS Motivated by #meuprofessordisse, physics students from the "Girls in Science" project at the Federal University of Rio Grande do Sul (UFRGS) collected comments heard by physics students across the country; they used an online anonymous survey. This group also brought hashtags to another dimension, producing images displaying what the faculty say to students in Brazil and sharing them through social media (Fig. 15.5). Here, hashtag activism was, somehow, institutionalized as the narratives were created through a federally funded project that aimed to foster the participation of young women in STEM fields.

Comments made by faculty in physics programs show sexism, homophobia, and abuse of power. When a faculty says they "thought women would not be able to do this activity," they explicitly doubt women's capacity in comparison to men. By doing so, they help mine their self-efficacy, which is an important factor for career choice and performance in science (Häussler and Hoffmann 2002).

Sexism, Racism, and Homophobia: Challenges in the Present Society

Considering what hashtag activism shows to be happening at universities, I believe it is fair to say hateful environment of society at large affects academic institutions. It would be naïve to think university spaces as safe heavens disconnected from the outside world. Even though a lot of what is discussed in academia stays within its walls, when it comes to peoples' behaviors and practices, it is impossible to dissociate faculty, students, and employees from their lives off campus. At least in this way, physics departments are not isolated from the rest of the world.

Women face, worldwide, the fear of being raped, just for being women. This hateful and extreme violence against women is still a problem in our society. When a physics professor says to a student "You are still going to be raped," this professor is perpetuating the misogyny we deal with on a daily basis. When this happens in a

I thought women would not be able to do this activity

You are girls, you are slower. If you didn't understand, I'm not going to explain it to you.

Women only get into college to find a husband

Turing is the mother of computer science

Depending on me, you won't graduate
*threat after a going out invitation

I need two Master students, an intelligent guy and a pretty girl to carry my books and to serve me coffee

Fig. 15.5 Images created to social media distribution displaying comments made by STEM faculty across the country

classroom, a place where we are supposed to feel safe, it can affect our ability to learn science content. This violence in a physics classroom can also interfere in a student's self-image toward becoming a scientist. How can a woman, surrounded by male colleagues and faculty, walk around and interact with all those men when she listens from one of them that she is going to be raped, one of the most severe violence a person can suffer?

#MyProfessorSaid shows us there are professors in Brazil who have been explicitly saying "physics is for men." This is not a subtle message; this is a clear message to any woman in a physics department that they do not belong there. Moreover, professors help perpetuating the image of women as people who exist to be pretty for the male gaze and who lack intellectual capacity. By saying "you are as beautiful as stupid" or "I will give you an easier exam because you are a woman," professors reinforce those ideas. If a woman wants to be in physics, the only path she is allowed to take is to be a high school teacher, which is thought to be easier and less of a career; according to what a secret professor said, "a physics teaching degree is for women who fail in a physics research degree." These secret professors bring to the classroom, in comments that may sound harmless, gender-negative stereotypes, "you are a woman; you should know how to cook," and career prejudice.

Physics professors exposed in #MyProfessorSaid send messages that constitute a public sphere denial form of oppression; they say to women not only where "their place" is but send a constant reminder that physics *is not* where they should be. These statements help keeping women out of public conversations about science, technology, and policies that involve this type of knowledge. Although women can now vote, they cannot fully participate in political decisions where scientific knowledge is required. Even when the discussion is about reproductive rights of women, the conversation, supposedly based on science, is led by men.

The stories brought by #MyProfessorSaid counter a dominant discourse that women do not like physics, that they are overly sensitive to "innocent jokes," and that students in the same classroom are given equal educational opportunities.

The vast majority of hashtags found use "professor" indistinctively of gender; however, there were a few that are explicitly marked as "professora," indicating the comment is made by a woman. When we look at those comments, the intersections of race and class become evident as privilege of White middle-class or upper middle-class women is exposed through racist and elitist remarks. Black and low-income students encounter a hostile environment in physics departments, according to the stories told by #MyProfessorSaid. The image of Black people as being low achievers and having less intellectual capacities is used to justify oppression in the form of racism. When a female professor asks physics students if they "want Black colleagues that know nothing" sharing that same space with them, she is saying non-Black students are more intelligent and deserve to be in that classroom, while a Black student should not be allowed entering that space. Moreover, she is influencing students to be against the presence of Black colleagues in physics; she is multiplying the chances a Black student will enter a racist environment when going into physics.

Brazilian physics students reveal how our physics departments are a site for all sorts of oppression. They show us that depending where one come from, geographically, they will not be seen as part of that community. At least one student has listened they "need to study more because of where [they] were born," suggesting people from certain parts of the country are less capable of following a physics course and disseminating oppression in the form of negative geographical stereotypes. The very fact a professor thinks that of their students may hinder students' learning in the classroom. There are even students who are advised to give up of physics and change careers, "If I were you, I would change major." Recognition from instructors is one factor in measuring one's self-efficacy in physics (Sawtelle et al. 2012); the less a student feels their teacher recognizes their ability to do physics, the less the said student will believe they can do physics.

Even common human respect of others' pain, as in the moment a family member passes away, seems to be forgotten in some Brazilian physics departments. Through #MyProfessorSaid, a student reports their professor's reaction when hearing about the death of the student's grandmother: "Did your grandma die? Bring me the death certificate; you will fail anyway." If this cold human being is the image one associates with being a physicist, maybe one will not want to be identified with this.

Continuing with the analysis of Brazilian physics departments through #MyProfessorSaid, we can see a lack of comments addressing sexual minorities. That can be read as a result of the severe underrepresentation of sexual minorities in physics. Brazil is a hostile country for LGBT+ population: it is "the deadliest country for sexual minorities" (Jacobs 2016). Because sexual minorities are not necessarily visible minorities, it is not a surprise to see or identify fewer of us in physics departments. That lack of visibility is not a constraint for professors to make homophobic remarks, though.

In the form of a homophobic joke, a professor said "Turing is the mother of computer science," making reference to Alan Turing a scientist who is known to be gay. This is a clear situation of microaggression to homosexual students; by making fun of a notable scientist's sexuality, this professor can send the message that no matter how brilliant a student is or how big their contribution to science is, they are going to be mocked for their nonnormative sexual orientation, they are going to be reminded they do not socially perform like a scientist should.

A homosexual physics student who walks around the Federal University of Rio de Janeiro campus may find a wall painting where it reads "Death to gays at UFRJ." These open threats help keeping sexual minorities in ghettoized occupations, not being fully participants in our society. For the few sexual minorities who enter STEM courses, some practical academic situations can be discouraging. A STEM undergraduate student reveal through social media the hurdles he faces just to submit a paper abstract into an online system (Fig. 15.6), he says:

> Great things about being trans in the academy: I'm here with my abstract ready to submit to the system but I haven't done it yet because no one has answered my emails about being allowed to use my real name in the system.

This hostile environment sexual minority people experience in academia is not a Brazilian phenomenon; it happens around the world. Diana Bilimoria and Abigail Stewart discuss the academic climate for LGBT+ faculty in science and engineering in the United States and show a negative climate can have career consequences such as "explicit exclusions from opportunities" (2009, p. 96). Louise Mayor, in a *Physics World* in-depth report, exposed intolerance toward sexual minorities at CERN particle-physics laboratory in Switzerland (2016). One year before that, the

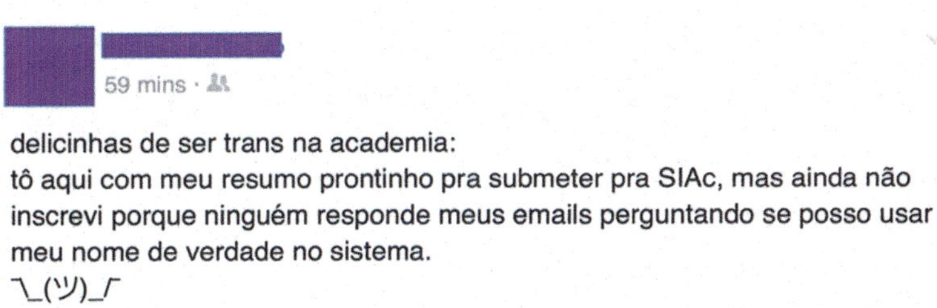

Fig. 15.6 A trans student uses social media to share his struggle to submit an abstract

American Physical Society (APS) ad hoc committee on LGBT issues (C-LGBT) produced LGBT Climate in Physics, a report that reviews the status of LGBT physicists to assess the barriers to full inclusion within the physics community around the world.

APS report revealed there is a degree of discrimination felt by sexual minorities, where gender-nonconforming people are more uncomfortable at their university departments and workplace and LGBT men face the least discomfort, followed by LGBT women. This was also reported for observing and experiencing harassment (Atherton et al. 2016). These data show the intersections of gender and sexuality make LGBT+ women experience oppression more than LGBT+ men. The situation is worst for gender-nonconforming people. These findings are along with Shayle Matsuda's work when discussing the hurdles faced by trans researchers in science (2015). In 2013, Out in Science, Technology, Engineering, and Mathematics conducted the first Queer in STEM survey (Out in science 2016). It was an in-depth look at the experiences of sexual minority professionals working in STEM fields. The survey was answered by more than 1400 people in a worldwide research, with larger participation of people from the United States and Canada. At the time, the Queer in STEM revealed more than 40% of sexual minorities in STEM have not disclosed their sexual identity to colleagues, coworkers, or students, even if they are totally "out of the closet" at home. Another finding was that STEM environments with more even representation of men and women and with colleagues and employers that openly support sexual minorities help people being more likely to be open about their identities. A second edition of this survey was conducted in December 2016, and its results are not out by the time of this publication. Overall, when it comes to sexual minorities in STEM fields, we still lack of research, particularly those looking at LGBT+ students' experiences. In this direction, analyzing hashtag activism is a powerful tool to access students' perspectives.

Finally, it is interesting to note a particular feature of hashtag activism in Brazil, its mix and merging of online and offline activism, particularly by writing post or printing what was first posted online. It looks like the timeline goes as a movement emerging online, connecting people from different locations, and then a counterstory is developed, and local communities bring the conversation back to include, expose, and face the perpetrators of the aggressions. This story timeline strengthens the narrative and protects the identity of students who revealed the comments in the first place. When hashtags come back to local institution on their walls, there is a greater level of anonymity.

15.4 Conclusions and Recommendations

Implications

The discussion brought in this paper has implications for the Physics Education Research (PER) community, for our school and academic practices, and for physics classrooms. It is important to PER people to start thinking about how race, gender, and sexuality may play a role in our research; these social constructs and identities need to be taken into account when studies are designed. By assuming students', faculty's, and teachers' identities do not affect the results of our studies might actually hinder the quality of our research. In addition, PER need to incorporate qualitative approaches and methodologies that allow for individual characteristics often lost in traditional quantitative approaches and to be aware of new sites of research such as social media.

Educators may have access to up-to-date PER and to learn all about innovative teaching strategies, but if none of these address the (not so micro) aggressions women, people of color, and sexual minorities face in physics departments, these resources will not be effective. We need to promote inclusive practices for racial, gender, and sexual minorities in our school, college, and universities. It is urgent that we work to accommodate name changes in publications or online submission systems, for example. Promoting the creation of interest/support groups for racial, gender, and sexual minorities is also something that can be done in science education community.

Changes can start even when selecting materials (e.g., textbooks) to work; it is important to look for materials that do no reinforce racial, gender, and sexual stereotypes. People need to (re)think in which way language, comments, and jokes made in class are promoting prejudice, racism, sexism, and/or LGBTphobia. Finally, we have to discuss the participation of racial, gender, and sexual minorities in physics. The following are selected resources that can help addressing some of these implications.

Selected Resources

- *National Organization of Gay and Lesbian Scientists and Technical Professionals* (NOGLSTP), a professional society that educates and advocates for lesbian, gay, bisexual, transgender, and queer students and professionals in science, technology, engineering, and mathematics. They produced a material "Queer scientists of historical note" that can be helpful for the classroom. In addition, they have a scholarship program for people who identify as LGBT+ http://www.noglstp.org.
- *LGBT Climate in Physics* report reviews the status of LGBT physicists to assess the barriers to full inclusion within the physics community. The document is available online at http://www.aps.org/programs/lgbt/upload/LGBTClimateInPhysicsReport.pdf.
- *LGBT+ physicists* is a website for and from LGBT+ and allies in physics, http://lgbtphysicists.org.
- *The American Physical Society page on LGBT Physicists*, http://www.aps.org/programs/lgbt.
- *American Institute of Physics* Teaching Guides on Women and Minorities in the Physical Sciences are valuable resources for the classroom, with full lesson plans available, https://www.aip.org/history-programs/physics-history/teaching-guides-women-minorities.

- *Out in Science, Technology, Engineering, and Mathematics* (oSTEM) is a support network which has been holding conferences targeted to LGBT+ people in STEM, http://www.ostem.org.
- *Prisma* is a recently formed group that aims to promote activities and discussions in order to transform the Federal University of Minas Gerais' Instituto de Ciências Exatas (ICEx) into a safe and welcoming environment for LGBT community, http://www.facebook.com/Prisma-603587806483371.
- *Negras e negros nas ciências* is a Brazilian community for and of Black people in science and others interested in supporting people of color in STEM, http://www.facebook.com/groups/negrasenegrosnasciencias.

15.5 Final Remarks

The analysis of a new era of activism through social media can be helpful for unveiling oppressive environments in academia. Specifically, this study shows the oppressive climate for gender, racial, and sexual minority Brazilian students in STEM.

The intersection of race, gender, and sexual orientation hinders the social integration of Black women and sexual minorities in academia and exposes the racism, elitism, and LGBTphobia among students and faculty in Brazilian STEM departments. There are no possibilities to shift the racialized gender experiences Black women face to solely gender experiences. Similarly, we cannot isolate sexual minorities' experiences in STEM without looking at the intersections of race and gender.

The performance expected by the scientific community for those entering STEM fields is the one that mimics this predominantly White, male, and heterosexual environment. Black women and sexual minorities embody precisely the opposite. These groups might face more obstacles to achieve the recognition component of scientific identity – recognizing oneself and being recognized by others.

Experiences of surprised looks from colleagues and teachers, differential treatment, and being outnumbered in a classroom send to young women, people of color, and sexual minorities the message they are out of place. Adding to this is an absence of representation of racial, gender, and sexual minority scientists in the media, textbooks, and lesson plans. Collectively, these microaggressions create an environment that teaches underrepresented minority youth that they do not belong in science and are not welcomed to pursue careers in STEM. As I have said in previous work, microaggressions at educational settings are a powerful tool to enculturate students in a stratified society and its systems of power and knowledge (Rosa 2013). Physics departments and school administrators need to address racist, sexist, and homophobic comment and behaviors. It is baffling the "secret professors" in this analysis did not face legal consequences for their acts.

References

Atherton, T., Barthelemy, R., Deconinck, W., Falk, M., Garmon, S., Long, E., Plisch, M., Simmons, E., Reeves, K. (2016). LGBT Climate in Physics: Building an Inclusive Community. American Physical Society, College Park, MD.

Aquino, Y. (2016, June 20). Após ato racist e homofóbico, UnB condena ofensas e vai apurar caso. Agência Brasil. Retrieved from http://agenciabrasil.ebc.com.br/geral/noticia/2016-06/universidade-de-brasilia-condena-ofensas-homofobicas-e-racistas

Berridge, S., & Portwood-Stacer, L. (2015). Feminism, hashtags and violence against women and girls. Feminist Media Studies, 15, 341–338.

Bilimoria, D., & Stewart, A. (2009). Don't ask, don't tell: the academic climate for lesbian, gay, bisexual, and transgender faculty in science and engineering. NWSA Journal, 21, 85–103.

Bonilla, Y., & Rosa, J. (2015). #Ferguson: digital protest, hashtag ethnography, and the racial politics of social media in the United States. American Ethnologist, 42, 4–17. doi:https://doi.org/10.1111/amet.12112

Butler, J. (1990). Gender trouble: feminism and the subversion of identity. New York: Routledge.

Carlone, H. B., & Johnson, A. (2007). Understanding the science experiences of successful women of color: science identity as an analytic lens. Journal of Research in Science Teaching, 44, 1187–1218.

Collins, P. (2000): Black feminist thought: Knowledge, consciousness, and the politics of empowerment. New York: Routledge.

Crenshaw, K. (1989). Demarginalizing the intersection of race and sex: a Black feminist critique of antidiscrimination doctrine, feminist theory and antiracist politics. University of Chicago Legal Forum, 1989, 139–167.

Crenshaw, K. (1993). Mapping the margins: intersectionality, identity politics, and the violence against women of color. Stanford Law Review, 43, 1241–1299.

Gauthier, J. (2016, June 14). Professores homossexuais são assassinados e corpos são queimados no interior da Bahia. Correio 24 horas. Retrieved from http://blogs.correio24horas.com.br/mesalte/professores-homossexuais-sao-assassinados-e-corpos-sao-queimados-no-interior-da-bahia/

Gerbaudo, P. (2012). Tweets and the streets: social media and contemporary activism. London: Pluto Press.

Häussler, P., & Hoffmann, L. (2002). An intervention study to enhance girls' interest, self-concept, and achievement in physics classes. Journal of Research in Science Teaching, 39, 870–888.

Hunt, R. (2016, June 13). The Orlando shooting grew out of everyday homophobia – we cannot be complacent. The Telegraph. Retrieved from http://www.telegraph.co.uk/news/2016/06/13/the-orlando-shooting-grew-out-of-everyday-homophobia%2D%2Dwe-cannot/

Jacobs, A. (2016, July 5). Brazil is confronting an epidemic of anti-gay violence. The New York Times. Retrieved from https://www.nytimes.com/2016/07/06/world/americas/brazil-anti-gay-violence.html

Lave, J., & Wenger, E. (1998). Communities of practice: learning, meaning, and identity. New York: Cambridge University Press.

Martin, M. (2016, July 7). Assassinato de estudante negro e gay no Rio escancara intolerância na universidade. El Pais Brasil. Retrieved from http://brasil.elpais.com/brasil/2016/07/05/politica/1467723193_955040.html

Matsuda, S. (2015, December 17). Trans researchers are struggling to stay in science. That has to change. Wired. Retrieved from https://www.wired.com/2015/12/science-needs-to-do-a-better-supporting-trans-scientists/

Mayor, L. (2016, March 3). Where people and particles collide. Physics World. Retrieved from http://physicsworld.com/cws/article/indepth/2016/mar/03/where-people-and-particles-collide

Open letter (2016, July 2). Students at Instituto Federal de Educação Ciência e Tecnologia de Goiás (IFG). Retrieved from http://www.facebook.com/IFGocupado/photos/a.196868690656190.1073741828.196867317322994/296083437401381

Out in Science, Technology, Engineering, and Mathematics. (2016, June 15). Introducing Queer in STEM 2.0, 2016. Retrieved from http://www.queerstem.org/2016/06/introducing-queer-in-stem-20.html
Pierce, C. (1978). Television and Education. Beverly Hills, CA: Sage.
Pierce, C., Carew, J., Pierce-Gonzalez, D., & Wills, D. (1977). An experiment in racism: TV commercials. Education and Urban Society, 10, 61–87.
Richard Delgado, (1989) Storytelling for Oppositionists and Others: A Plea for Narrative. Michigan Law Review 87 (8):2411
Rosa, K. (2013). Gender, ethnicity, and physics education: understanding how Black women build their identities as scientists. (Doctoral dissertation). Retrieved from http://hdl.handle.net/10022/AC:P:18782
Rosa, K., & Mensah, F. (2016). Educational pathways of Black women physicists: Stories of experiencing and overcoming obstacles in life. Phys. Rev. Phys. Educ. Res., 12, 020113 doi: https://doi.org/10.1103/PhysRevPhysEducRes.12.020113
Sawtelle, V., Brewe, E., Goertzen, R. M., & Kramer, L. H. (2012). Identifying events that impact self-efficacy in physics learning. Phys. Rev. ST Phys. Educ. Res., 8, 020111 doi: https://doi.org/10.1103/PhysRevSTPER.8.020111
Stanton, E. C., Anthony, S. B., & Gage, M. J. (1889). History of woman suffrage, 2nd ed., vol. 1, Rochester, NY: Charles Mann.
Yang, G. (2016). Narrative agency in hashtag activism: the case of #BlackLivesMatter. Media and Communication, 4, 13–17 doi: https://doi.org/10.17645/mac.v4i4.692

Katemari Rosa is a Professor of Physics at the Federal University of Bahia, in Brazil. She identifies as a Black Latina bisexual woman. She works primarily with preservice physics teachers and is committed to the teaching of science for social justice. Her research focuses on the intersections of gender, race, and ethnicity in physics education. She is a member of the Brazilian Physics Society and the American Association of Physics Teachers.

Chapter 16
Diversity, Human Rights and Physics Education: Theoretical Perspectives and Critical Awareness

Tanja Tajmel

Abstract This chapter deals with diversity in physics education through a discourse analytical lens by examining the discourse on diversity from different perspectives. One perspective is the utilitarian one, which sees diversity as a human resource that should not remain untapped. Thus, physics education should promote those, who, so far, have been underrepresented in physics and science careers. The other perspective is the emancipatory-humanistic one, which considers disparities in performance and careers as indicator for social inequalities resulting in limited access to physics education. From this perspective, physics education itself might contribute to social inequality as it reproduces certain patterns of inequality. The two perspectives provide different arguments for the promotion of diversity in science education and have huge influence on the common perception and justification of physics education itself. Is the primary task of physics education to "produce" economically exploitable scientific workforce, or is physics education of importance for the empowerment of the individual, and, thus, independent from economic exploitability? What does it mean for educators to promote or consider diversity in physics education in this respect? As physics teachers and science educators are agents in the process of scientific socialisation, a critical awareness of teachers and education researchers amongst diversity becomes increasingly relevant. Considering diversity in the common conceptual delineation of diversity—categorising individuals by certain characteristics from an essentialist standpoint—bears the risk of tokenising, stereotyping, "othering" and discrimination. The human rights perspective and the right to STEM education provide an approach towards a critical understanding of diversity and a framework for the empowerment of the individual.

T. Tajmel (✉)
Centre for Engineering in Society, Concordia University, Montreal, QC, Canada

M. Pietrocola (ed.), *Upgrading Physics Education to Meet the Needs of Society*,
https://doi.org/10.1007/978-3-319-96163-7_16

16.1 Introduction

Without doubt, considering diversity on all levels of society is positive. It is one step forward towards meeting injustice und inequality. Looking at different statements addressing diversity, it becomes evident that distinct interests are connected with the goal to promote diversity, especially in the sciences and in science education. Citing the website of Nestlé, one of the biggest global players in food industry, makes clear that the motivation to promote diversity grounds at least on two different motives:

> Yes, we believe we have a social responsibility to promote diversity, but equally we know it sharpens our performance and gives us an advantage over our competitors. (http://www.nestle.com/jobs/your-career-at-nestle/your-work-life)

Another global player, Monsanto, stated similarly:

> We search for the brightest talent, regardless of geography, gender, skin color, age or disability. (http://www.monsanto.com/global/in/careers/pages/diversity.aspx)

Diversity means increasing chances and success in two ways: (1) for the individuals who are empowered to choose freely and equally the way in which they want to live and (2) for the competitors and the competition within the global market. Promoting diversity amongst the so-called human capital (see Keeley 2007) means increasing the chances for an efficient workforce. However, regarding science and technology, such arguments are not only of interest for the sake of the wellbeing of all people. Scientific progress is of utmost relevance for the efficiency of industrialised production of a huge variety of products—be it coffee, seeds, sneakers, oil, meat, electronic devices or weapons. Due to these different interests in promoting diversity, a *critical* awareness on these topics becomes increasingly relevant. Science educators, such as physics teachers and physics education researchers, are the keys. They can consider themselves both as trainers of tomorrow's workforce and as mentors for critical thinking, self-confident and empowered individuals.

By referring to theoretical perspectives of sociology and human rights, the present chapter aims at raising awareness and critical thinking in the field of physics education and physics education research.

- When did *diversity* become of interest in science and physics education?
- Why should physics education address diversity? (*Economic reasons? Emancipatory reasons?*)
- On which theoretical basis can we ground the argumentation?

16.2 Physics Education as a Social Field

In order to develop critical awareness on the relation of physics education and social power, one approach is considering physics education as a social field in the sense of Bourdieu (1987), where actors hold certain positions determined by the rules of the field. From this perspective, the specific "culture of physics" provides norms to include or exclude individuals from certain positions in this field. In this context, valuable approaches on science education as social field and on science as specific culture have been provided by Aikenhead (1996, 2006), Costa (1995), Harding (1991, 2006), Lemke (1990, 2001, 2011), Roth and Lee (2002), Tobin (2009), Wegerif et al. (2013), Van Eijck (2013), and Hussénius (2014), amongst others. From a critical standpoint, the following key questions arise: Who does define the goals of science education? Who does benefit from (in a certain way) scientifically educated people? How can educators increase the critical thinking of students, teachers and science education researchers? How can the social field of science open up and become inclusive? As physics education is the link between society and scientific progresses, science teachers and science education researchers are multipliers and, therefore, "agents of socialisation". As Lemke states:

> It is not surprising that those who succeed in science tend to be like those who define the "appropriate" way to talk science: male rather than female, white rather than black, middle- and upper-middle class, native English-speakers, standard dialect speakers, committed to the values of North European middle-class culture (emotional control, orderliness, rationalism, achievement, punctuality, social hierarchy, etc.). (Lemke 1990: 139)

Clearly, the characteristics of the specific culture of science limit the access to science for certain students. In the words of Aikenhead:

> [S]cience educators, Western and non-Western, need to recognize the inherent border crossings between students' life-world subcultures and the subculture of science, and that we need to develop curriculum and instruction with these border crossings explicitly in mind, before the science curriculum can be accessible to most students. (Aikenhead 1996: 2)

In this way, cultural characteristics work as discriminating factors hampering the access to science and, in consequence, increasing social inequality (Harding 2006; Tajmel 2017).

16.3 "Othering" in Physics Education

Othering, as the term first used in postcolonial writing in a systematic way (Said 1978), means categorising and labelling human beings in order to structure society and to constitute hegemonic order as common sense and *normality*. Sociologically, the construction of the *other* allows a distinction to the *we*, that is, the normality, which is, in general, not defined or described more precisely. Talking about diversity means talking about differences and thus different categories. Especially when talking about *migrants* or *ethnic origin*, *othering* might occur, as such labelling carries postcolonial characteristics. The label *migrant* is, in its common sense, accompanied by other labels: poor educational background, language deficits, non-Christian (Muslim), black, etc.

In the social field of physics and physics education, processes of *othering* occur when certain characteristics are being considered appropriate or not appropriate for physics. Some gain access to physics and become part of the culture of physics, while others do not. Thus, the relevant questions to be asked are as follows: What is being considered as *normality* in physics education? Who is considered to have suitable preconditions, to be or not to be *scientifically interested*, to be or not to be *talented* or to talk science *the right way* (see Lemke 1990)?

16.4 Diversity from Utilitarian Perspective

In order to understand the utilitarian perspective for promoting diversity, underprivileged groups and so-called low performers in science education, one has to look back into an era, when science education has been at the edge of becoming a socially and politically highly relevant topic.

In 1957, when the Soviet Union launched the first satellite *Sputnik*, the whole Western world was shocked. This event marked a milestone in the scientific and technological competition between the two superpowers of that time. From that on, science education became a national interest. Under US President Eisenhower, the *National Defense Education Act* was passed. Mathematics and science became important as school subjects, and an immense number of actions have been taken for promoting science education. Especially female students, who demonstrated poor performance in mathematics and sciences, were regarded as the target group to be supported and encouraged in science. At the same time, the women's movement demanded more rights for women. The promotion activities for women can thus be seen as motivated by emancipatory reasons as well as by reasons of technological competition.

In 2000, the so-called PISA-Shock triggered promotion activities for science education, which are comparable with the initiatives taken after Sputnik. However, there exists a relevant difference: Sputnik occurred as a result of technological and scientific progress, materialised in the form of a satellite. PISA, on the other hand, was planned by the OECD, the Organisation for Economic Co-operation and Development (OECD 1999). The reference point were performance targets of students set by the OECD. One key outcome was that there exist so-called low-performing students, and certain factors were identified as being responsible for the low performance (OECD 2006). These factors are (amongst others) gender, family, living area, migration background, language and socio-economic background.

The genesis of the social and political discourse on science education as triggered by PISA laid the foundations for seeking the relevant deficits and faults in the person's identity itself. Statements like "Students perform poorly *because* of the migrant background or *because* of their socio-economic status or *because* of their family language" clearly locate the deficits in the students themselves and in their families rather than in the social or educational processes that *make* certain groups performing poorly.

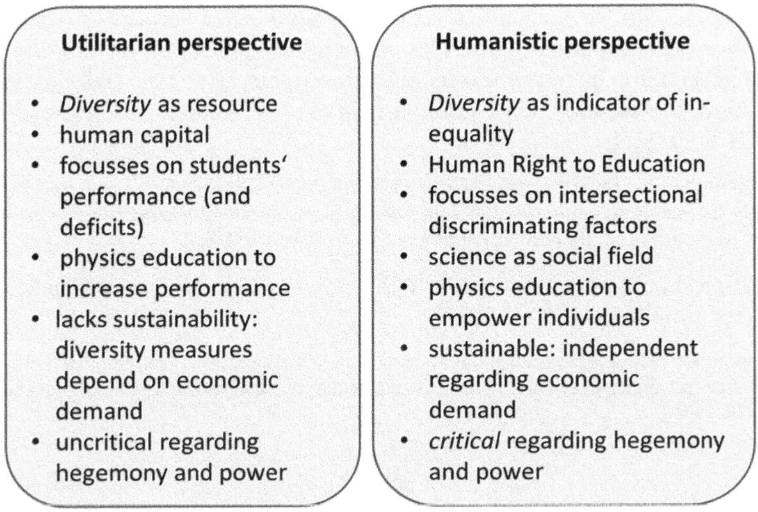

Fig. 16.1 Juxtaposition of the utilitarian and the humanistic perspective on diversity

16.5 The Human Rights Perspective

With regard to diversity, maybe the most important characteristic of the declaration of the human rights is that human beings are not categorised or labelled at all. The declaration states discrimination as strictly forbidden and names discrimination on grounds of class, race, gender, religion and body, amongst others. From this perspective, in order to counteract discrimination, first of all the *reasons for discrimination* have to be analysed rather than the different characteristics of certain human groups, by which *diversity* is commonly conceptualised. Figure 16.1 shows the juxtaposition of the utilitarian and the humanistic perspectives on diversity.

Human rights are rights of self-determination under the terms of equality of rights for all. Article 26 of the Universal Declaration of Human Rights 1948 (UDHR), the Convention against Discrimination in Education 1960 (CADE) and Art. 13 of the Covenant on Economic, Social and Cultural Rights 1966 (ICESCR[1]) are the most important and fundamental internationally recognised provisions stipulating the "right to education" without discrimination or exclusion. The substance of the right to education aims (1) at the development of the human personality and consciousness of human dignity and fostering respect of human rights and (2) at enabling individuals to play an active and useful role in the society (Starl 2009: 19f). In the framework of human rights, *human dignity* is an end in itself. As rights of equal freedom, human rights are the opposite of privileges, because human rights do not refer to specific characters or qualifications, but to being human as the sole ground. This is exactly what is meant by the universality of human rights: *every*

[1]International Covenant on Economic, Social and Cultural Rights (ICESCR 1966)

human being is entitled to all and the same (human) rights because of being human and for the sake of human dignity. This implicitly excludes any form of discrimination logically. Individuals are bearers of human rights, which are obliging the state (s). The right to education has a solid basis in the international and regional human rights law in general:

> [A]s an empowerment right, education is the primary vehicle by which economically and socially marginalized adults and children can lift themselves out of poverty and obtain the means to participate fully in their communities. (13 ICESCR 1966)

Education should be empowerment to the individual in order to participate fully in this society. And further:

> But the importance of education is not just practical: a well-educated, enlightened and active mind, able to wander freely and widely, is one of the joys and rewards of human existence. (CESCR 1999)

16.6 The 4-A Scheme

The right to education is being structured by four essential features (see Tomaševski 2006): *availability*, *accessibility*, *acceptability* and *adaptability* (Table 16.1). This so-called 4-A scheme was suggested by Katarina Tomaševski, the former UN Special Rapporteur on education, and formulated by the CESCR in the General Comment 13 (CESCR 1999, para. 6). The 4-A scheme is both a threshold and a tool for the evaluation of the compliance of a respective education system with the right to education.

Availability

Functioning educational institutions have to be available in sufficient quantity. Functionality requires, amongst other features (well and adequately), trained teachers receiving domestically competitive salaries and teaching materials. According to the terms of economic capacity and needs of a country, some will

Table 16.1 The 4-A scheme (CESCR 1999; Tomaševski 2006; Starl 2009; Tajmel 2017): structure and essential features of the right to education

Characteristics	4-A Scheme (CESCR 1999; Tomaševski 2006; Starl 2009; Tajmel 2017)
Availability	Free education, infrastructure, buildings, books, materials, qualified staff and teacher trainings
Accessibility	Equal access, physically and economically, regardless of socio-economic background, language, religion, ethnicity
Acceptability	Curricula and lessons should be unbiased, free of discrimination and relevant to all students
Adaptability	Flexible educational programs according to changes in society

also require facilities such as a library, computer facilities and information technology (CESCR 1999, para. 6 (a)).

Article 13 (2) (b) ICESCR states that secondary as well as vocational education shall be "generally available". This signifies that secondary education is not dependent on a student's apparent capacity, ability or merit. It means further that secondary education will be provided in a way that it is available on the same basis to all, i.e. without any discrimination.

Accessibility

Educational in all its forms, institutions and programmes must be accessible to everyone without discrimination. Accessibility has three interrelated dimensions (CESCR 1999, para. 6 (b)):

- Non-discrimination—education must be accessible to all, especially to the most vulnerable groups.
- Physical accessibility—education has to be "[...] within safe physical reach, either by attendance at some reasonably convenient geographic location (e.g. a neighbourhood school) or via modern technology (e.g. access to a 'distance learning' programme)" (CESCR 1999, para 6 (b)).
- Economic accessibility—education has to be affordable to all. This specific dimension of accessibility is often underestimated or misunderstood. Firstly, economic accessibility requires that admission to school shall be free. In most of the European countries, most of the primary and secondary, also post-compulsory and vocational, schools are free. However, this might be misleading in a sense that there also arise enormous costs apart from school fees. Hence, (relative) poverty, social or economic status as well as birth must not prevent individuals from access to all forms and levels of education. Barriers in economic accessibility include also indirect costs and opportunity costs of education (see Tomaševski 2006: 27). However, the pecuniary side is only one aspect of the economic accessibility. Accessibility also means to have the de facto opportunity to access all forms and levels of education. This has also something to do with the *social* and *cultural capital* of a person (Bourdieu 1987, 1991): "In respect to social class and economic standing the considered education systems [e.g. of Germany, Austria, Turkey] clearly fail. They tend to reproduce existing social and economic inequality as we learn from a long list of educational research [..]. Regrettably, we can say that the education systems even corroborate inequality in the long-run. Limited economic access to education is not only unfair. It is a human rights violation as it is discriminatory on the grounds of social origin and property to omit effective compensation or affirmative action". (Starl 2009: 23)

Acceptability

> The form and substance of education, including curricula and teaching methods, have to be acceptable (e.g. relevant, culturally appropriate and of good quality) to students and … [their] parents …. (CESCR 1999, para. 6 (c))

More particularly, the meaning of *culturally appropriate* reveals an obligation of the state to deliver secondary and vocational education, which (1) is culturally sensitive towards the diversity of students by appropriate science lessons, curricula and equipment, (2) promotes the mutual understanding of cultures and (3) enables students to acquire the necessary knowledge and skills for their self-reliance. The term *cultural appropriateness* indicates that education has to take into account the cultural background of students by providing equal opportunities instead of putting pressure to assimilate.

Adaptability

The principle of adaptability requires the education to be flexible in the sense that it can adapt to the needs of changing societies and respond to the needs of students within their diverse social and cultural settings. The principle of adaptability not only applies to the education system as a whole but also to all forms and levels of education. Moreover, it applies to all of its components, e.g. equipment of schools or universities, teaching material, curricula and the knowledge and skills of the teaching staff. Neglecting the different needs of students in the supply and maintenance of education disadvantages certain groups by limiting their access and acceptance and, therefore, privileges others.

16.7 Barriers to the Access to Physics Education

Barriers are *exclusion mechanisms* instead of disadvantages rooted in the identity of the individual and, on the one hand, may be intentionally set or constructed. On the other hand, barriers can also exist due to historical developments and the lack of socially necessary adaptation. Particularly, barriers in the access to education have often the character of suggesting individual responsibilities for underperformance, underachievement, limited access or whatever lack of opportunities. This fact builds up a barrier by itself. Barriers may prevent either a majority or a minority from full access to education or from the opportunity to personal development. Any exclusion based on a prohibited ground as colour, "race", ethnicity, age, body, religion or gender (and others) is not compliant with the fundamental principle of human rights. This principle justifies affirmative action to guarantee equality. Individual disadvantages have to be compensated for by affirmative action to equalise opportunities. They are subject to discrimination when used as grounds for distinction or exclusion, no matter whether intended or resulting from provisions or policies.

However, barriers are hard to identify in most cases. Qualitative and quantitative indicators are needed to identify barriers, their consequences, their impact and the causalities between structures and procedures. *Achievement rates* are recognised indicators for the existence of barriers to the access to education. From this perspective, the disparities of students' performance are indicators for serious barriers, which hamper the access to education. Consequently, the focus of research lies on the processes hampering individuals in physics education rather than on the individuals themselves.

From the human rights perspective, the challenge is a sociocultural one: to develop and fundamentally change certain traditions and routines within the culture of physics and physics education, which privilege some (male, white, nonmigrants, speaking standard language) and discriminate others (female, black, migrant-background, speaking non-standard language).

Conceptual Delineation of "Barrier"

In the framework of the project PROMISE—Promotion of Migrants in Science Education—a barrier analysis has been performed based on the right to education (Tajmel and Starl 2005; Tajmel et al. 2009). On the basis of the problem analysis, we identified three types of barriers that impede access to education: *linguistic*, *cultural* and *institutional*. In our model, the socio-economic background of students is not considered as a barrier. Socio-economic differences are facts in every society, and open school systems make—or try to make—accessibility to primary and secondary education independent from one's economic status. The fact that the school system represents a certain social class and culture and, thus, predominantly addresses students with a certain cultural capital (which includes a certain social and economic background) is considered here as *cultural barrier*.

Linguistic barriers are considered as deficits of the educational processes rather than of the students or their families. Multilingual students and students with other languages than the standard language are generally disadvantaged in classes, since they have to communicate in a language which is neither their first nor best. Linguistic barriers can be understood as a barrier caused by insufficient consideration of the needs of second-language learners and multilingual students in physics education.

The cultural habitus of a school is indicated by the cultural and social ancestry of teachers and by the cultural capital (Bourdieu 1991) which the system demands from the students. The monocultural habitus is reflected by the depiction of subject-specific contents in schoolbooks and also by that of persons and objects in textbooks. They do not reflect the diversity of the society but, instead, a certain social class, language and culture. Based thereon, ethnic and gender differences are being constructed and co-constructed within the school. Raising consciousness and establishing a certain preparedness for change are very difficult, particularly in the field of physics due to the common assumption that physics and science are objective and value-neutral. Alike linguistic barriers, cultural barriers are not deficits of the students, but, rather, deficits of the physics classes and educational processes.

Institutional barriers are understood as barriers on which individual qualified teachers or teacher students do not have relevant influence. Examples for institutional barriers are teacher trainings, which insufficiently educate teachers in handling new challenges brought about by a linguistically and culturally heterogeneous society. The lack of diversity education amongst teachers leads to insufficient skills. Further institutional barriers can be identified in the standards, in the curricula, in the determination of class sizes and in the lack of conditions for teachers' co-operation.

16.8 Access to Physics Education: Intersectional Factors

The role of physics education in the context of the right to education can be summarised as follows:

- Science education and physics education are relevant parts of education. Consequently, every human being has the right to science education.
- The physics lesson is the institutionalised access to science education. Thus, the 4-A scheme can be applied to physics education.
- Physics and the physics lesson are social fields shaped by social practices (identity construction, cultural representation and institutional structure). These social practices can support or hamper the access to science education by privileging or discriminating certain groups or individuals.

By taking these facts into account, the key question for making physics education accessible to all students arises as follows: *By which means do physics education and the culture of physics enhance social selection processes?*

Taking into account *intersectional discrimination* (Crenshaw 1989), that is, multiple discrimination on different grounds and on different analytical levels, the access to physics education can be modelled by three main factors interacting in an intersectional way: (1) the structure of science education, (2) the culture of physics and (3) the identities, which are constructed or co-constructed within educational processes, e.g. in physics classes (Fig. 16.2).

16.9 "Simulated Othering": Towards a Critical Diversity Awareness in the Physics Classroom

Simulated othering (Tajmel 2009, 2017) is a method to sensitise teachers for processes of *othering*, which occur in the physics classroom and co-construct *identities* that *differ* from the normality (the *we*). Here, the method is being applied on the situation of the observation of a physical experiment.

The demonstration and observation of an experiment are essential elements of the physics classroom. Putting the observation into words affords specific language

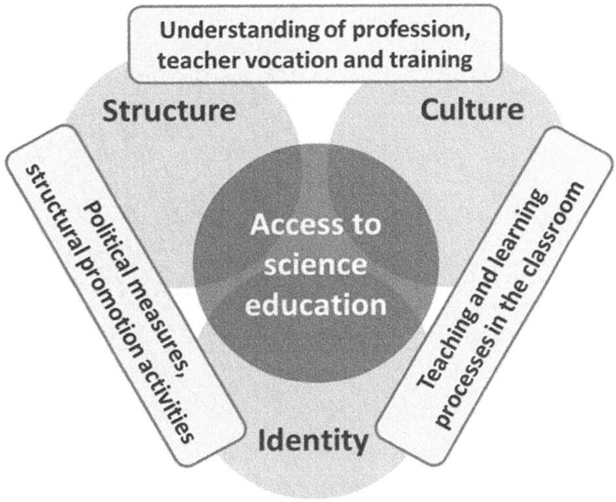

Fig. 16.2 Model of multifactorial determined access to science education (Tajmel 2017: 140)

skills, even when no technical terminology is required. Teachers try to ease the task by asking the students to describe the observation "in their own words". At the first impression, this task seems to meet all the different linguistic competencies of all students, as every student has his or her *own words* and every *own word* is allowed. However, unconsciously the teacher expects certain linguistic skills (own words, but in the standard language of the classroom) and a certain timespan, which the student needs for expressing the observation. Correspondingly, if a student does not have the required capacity of expression, the task is, to a certain extent, not fulfilled. Second-language learners need more time for expression, and the required *own words* are, in fact, words in their second, not in their first, language and will, thus, assumingly differ from the *own words* of a native speaker. This situation does not provide equal opportunities for all students as they do not have equal chances for interaction in the physics lesson (Fig. 16.3).

To sensitise teachers for this specific problem, teachers themselves are being put into the position of students by a teacher trainer and, therefore, *change their perspective*. The trainer conducts a physical experiment. The teachers are asked to observe the experiment, to describe the observation by 2–3 sentences and to write them down. The teachers are requested *not to use their native language but a second or foreign language*. Thus, the teachers become *others*. The language, which is required in this simulated physics lesson, is not their best or native language.

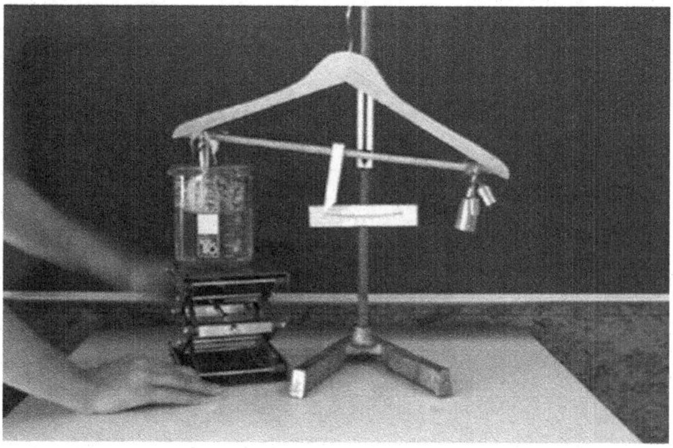

Fig. 16.3 Key experiment for exploring *simulated othering* (Tajmel 2009: 204)

Box. 1 The Task for the Teachers in Order to Simulate the Students' Position: "Simulated Othering" (Tajmel 2009: 205)

Observe the experiment.
Describe your observation.
Do not use your first language; use a foreign language.
Write your description down.
How do you feel?

Most of the teachers are surprised how difficult it is to put the observation into words in a foreign language. The teachers report to get "the feeling" of the situation of a student whose first language differs from the language of the science classroom. As the teachers individually experience linguistic barriers, their understanding of the situation and their motivation to change the situation increase. Most of the teachers express that they feel uncomfortable and embarrassed. They do not dare to read aloud. Three examples of teachers' statements (Tajmel 2017, original text in German, translated by the author) are as follows:

- "I cannot express precisely what I have observed". (MAVI008)
- "It's a shame how bad my language skills are. I feel helpless". (MAVI035)
- "I am embarrassed and wouldn't dare to read aloud". (MAVI127).

These statements indicate the experience of the individual as *socially positioned* by characteristics of the lesson (language, time) and *not* by characteristics of his or her personal attributes (e.g. observation skills). The psychological effect of *othering* is that individuals are embarrassed and their self-confidence decreases, as they find themselves in a weaker (inferior) position compared to those whose first language is the same as the lesson's language, the *normal language*. If a different lesson framework were normality, e.g. if the individuals could answer in their best language or if they got enough time and linguistic support, they would not feel embarrassed; they would not be *othered*. The experience of *othering* initiates a process of self-reflection on the teachers' position and on social structuring processes in the physics classroom.

16.10 Conclusion

Why should physics education deal with diversity? Educational processes are social processes. Students' assessments do not only reflect students' performance but also discrimination in students' access to education. In order to make physics education accessible to all students, the means by which physics education enhances social selection have to be carefully investigated. The utilitarian perspective does not provide a sufficiently critical approach to detect all barriers, whereas the sociocultural humanistic perspective is critical with respect to education and power. The human rights offer a profound framework to focus on the access to education and on empowerment of human beings rather than on students' deficits in performance. Each human being has the right to physics education. In this sense, respecting students' diversity means deconstructing *the other* (e.g. by *simulated othering*), opening the scientific culture and sustaining students in all imaginable different ways.

References

Aikenhead, G. (1996): Science Education: Border Crossing into the Subculture of Science. In: Studies in Science Education, 27: 1–52

Aikenhead, G. (2006): Science Education for Everyday Life. Evidence-Based Practice. Teachers College Press, Columbia University, New York.

Bourdieu, P. (1987): Der feine Unterschied. Kritik der gesellschaftlichen Urteilskraft, suhrkamp taschenbuch wissenschaft, Bd. 658. Suhrkamp, Frankfurt/Main.

Bourdieu, P. (1991): Language and symbolic power. Harvard University Press, Cambridge.

Committee on Economic, Social and Cultural Rights (CESCR), *General Comment No. 13, The Right to Education (Art.13)*, Twenty-first session, 8 December, 1999, available online at: http://www.unhchr.ch/tbs/doc.nsf/(symbol)/E.C.12.1999.10.En?OpenDocument (last checked: 11 June 2008).

Costa, V. (1995): When science is another world: Relationships between worlds of family, friends, school, and science. In: Science Education, 79: 313–333.

Crenshaw, K. (1989): Demarginalizing the Intersection of Race and Sex: A Black Feminist Critique of Antidiscrimination Doctrine, Feminist Theory and Antiracist Politics. In: The University of Chicago Legal Forum, 1989, 1.

Harding, S. (1991): *Whose Science? Whose Knowledge? Thinking from Women's Lives.* Cornell University Press, Ithaca, New York

Harding, S. (2006): *Science and Social Inequality. Feminist and Postcolonial Issues. Race and Gender in Science Studies.* University of Illinois Press, Urbana and Chicago.

Hussénius, A. (2014): Science education for all, some or just a few? Feminist and gender perspectives on science education: a special issue. In: Cultural Studies of Science Education, 9, 2: 255–262.

International Convenant on Economic, Social and Cultural Rights (ICESCR), Adopted and opened for signature, ratification and accession by General Assembly resolution 2200A (XXI) of 16 December 1966, available online at: https://www.ohchr.org/en/professionalinterest/pages/cescr.aspx (last checked: 11 December 2018).

Keeley, B. (2007): OECD Insights: Human capital. How what you know shapes your life. OECD Publishing, Paris.

Lemke, J. (1990): *Talking Science: Language, Learning and Values*. Language and Educational Processes. Ablex Publishing, Westport.

Lemke, J. (2011): *The secret identity of science education: masculine and politically conservative?* In: Cultural Studies of Science Education, **6**: 287–292.

Lemke, J. (2001): *Articulating communities: Sociocultural perspectives on science education*. In: Journal of Research in Science Teaching, **38**, 3: 296–316.

OECD (1999): *Measuring students knowledge and skills: A new framework for assessment. (Organisation for Economic Co-Operation and Development)*. OECD, Paris.

Organisation for Economic Co-Operation and Development (OECD) (ed.), *Where Immigrant Students Succeed – A Comparative Review of Performance and Engagement in PISA 2003*, OECD, Paris, 2006.

Roth, W.-M. & Lee, S. (2002): *Scientific literacy as collective praxis*. In: Public Understanding of Science, **11**, 1: 33–56.

Said, E. (1978): Orientalism. Vintage, New York.

Starl, K. (2009): *The Human Rights Approach to Science Education*. In: Tajmel, T. & Starl, K. [Eds.]: *Science Education Unlimited. Approaches to Equal Opportunities in Learning Science*, 19–36. Waxmann, Münster, New York.

Tajmel, T. (2009): *Does Migration Background Matter? Preparing Teachers for Cultural and Linguistic Diversity in the Science Classroom*. In: Tajmel, T. & Starl, K. [Eds.]: *Science Education Unlimited. Approaches to Equal Opportunities in Learning Science*, 201–214. Waxmann, Münster, New York.

Tajmel, T. (2017): Naturwissenschaftliche Bildung in der Migrationsgesellschaft. Grundzüge einer Reflexiven Physikdidaktik und kritisch-sprachbewussten Praxis. Springer VS. Wiesbaden, New York.

Tajmel, T. & Starl, K. (2005): *PROMISE - Promotion of Migrants in Science Education*. ETC Occasional Paper No. 18. URL: http://etc-graz.at/typo3/ index.php?id=74, last checked: 12/11/ 2016.

Tajmel, T., Starl, K. & Schön, L.-H. (2009): *Detect the Barriers and Leave Them Behind – Science Education in Culturally and Linguistically Diverse Classrooms*. In: Tajmel, T. & Starl, K. [Eds.]: Science Education Unlimited. Approaches to Equal Opportunities in Learning Science, 67–84. Waxmann, Münster, New York.

Tobin, K. (2009): *Difference as a resource for learning and enhancing science education*. In: Cultural Studies of Science Education, **4**, 4: 755–760.

Tomaševski, K., *Human Rights Obligations in Education: The 4-A Scheme*, Wolf Legal Publishers, Nijmegen, 2006.

Van Eijck, M. (2013): *Reflexivity and Diversity in Science Education Research in Europe: Towards Cultural Perspectives*. In: Mansour, N., Wegerif, R., Milne, C., Siry, C. & Mueller, M. P. [Hrsg.]: *Science Education for Diversity*, Cultural Studies of Science Education, 65–78. Springer, Heidelberg New York London.

Wegerif, R., Postlethwaite, K., Skinner, N., Mansour, N., Morgan, A. & Hetherington, L. (2013): *Dialogic Science Education for Diversity. Theory and Practice*. In: Mansour, N., Wegerif, R., Milne, C., Siry, C. & Mueller, M. P. [Hrsg.]: *Science Education for Diversity*, Cultural Studies of Science Education, 3–22. Springer, Heidelberg New York London.

Websites

http://www.monsanto.com/global/in/careers/pages/diversity.aspx (last checked 01/13/2017)
http://www.nestle.com/jobs/your-career-at-nestle/your-work-life (last checked 01/13/2017)

Printed by Printforce, the Netherlands